U.S. Conventional Prompt Global Strike

ISSUES FOR 2008 AND BEYOND

Committee on Conventional Prompt Global Strike Capability

Naval Studies Board

Division on Engineering and Physical Sciences

NATIONAL RESEARCH COUNCIL
OF THE NATIONAL ACADEMIES

THE NATIONAL ACADEMIES PRESS
Washington, D.C.
www.nap.edu

THE NATIONAL ACADEMIES PRESS 500 Fifth Street, N.W. Washington, DC 20001

NOTICE: The project that is the subject of this report was approved by the Governing Board of the National Research Council, whose members are drawn from the councils of the National Academy of Sciences, the National Academy of Engineering, and the Institute of Medicine. The members of the committee responsible for the report were chosen for their special competences and with regard for appropriate balance.

This study was supported by Contract No. N00014-05-G-0288, DO #16 between the National Academy of Sciences and the Department of the Navy. Any opinions, findings, conclusions, or recommendations expressed in this publication are those of the author(s) and do not necessarily reflect the views of the organizations or agencies that provided support for the project.

Cover: Images courtesy of the U.S. Navy.

International Standard Book Number-13: 978-0-309-11459-2
International Standard Book Number-10: 0-309-11459-4

Copies of this report are available from:

Naval Studies Board, National Research Council, The Keck Center of the National Academies, 500 Fifth Street, N.W., Room WS904, Washington, DC 20001; and

The National Academies Press, 500 Fifth Street, N.W., Lockbox 285, Washington, DC 20055; (800) 624-6242 or (202) 334-3313 (in the Washington metropolitan area); Internet, http://www.nap.edu.

Printed in the United States of America

THE NATIONAL ACADEMIES

Advisers to the Nation on Science, Engineering, and Medicine

The **National Academy of Sciences** is a private, nonprofit, self-perpetuating society of distinguished scholars engaged in scientific and engineering research, dedicated to the furtherance of science and technology and to their use for the general welfare. Upon the authority of the charter granted to it by the Congress in 1863, the Academy has a mandate that requires it to advise the federal government on scientific and technical matters. Dr. Ralph J. Cicerone is president of the National Academy of Sciences.

The **National Academy of Engineering** was established in 1964, under the charter of the National Academy of Sciences, as a parallel organization of outstanding engineers. It is autonomous in its administration and in the selection of its members, sharing with the National Academy of Sciences the responsibility for advising the federal government. The National Academy of Engineering also sponsors engineering programs aimed at meeting national needs, encourages education and research, and recognizes the superior achievements of engineers. Dr. Charles M. Vest is president of the National Academy of Engineering.

The **Institute of Medicine** was established in 1970 by the National Academy of Sciences to secure the services of eminent members of appropriate professions in the examination of policy matters pertaining to the health of the public. The Institute acts under the responsibility given to the National Academy of Sciences by its congressional charter to be an adviser to the federal government and, upon its own initiative, to identify issues of medical care, research, and education. Dr. Harvey V. Fineberg is president of the Institute of Medicine.

The **National Research Council** was organized by the National Academy of Sciences in 1916 to associate the broad community of science and technology with the Academy's purposes of furthering knowledge and advising the federal government. Functioning in accordance with general policies determined by the Academy, the Council has become the principal operating agency of both the National Academy of Sciences and the National Academy of Engineering in providing services to the government, the public, and the scientific and engineering communities. The Council is administered jointly by both Academies and the Institute of Medicine. Dr. Ralph J. Cicerone and Dr. Charles M. Vest are chair and vice chair, respectively, of the National Research Council.

www.national-academies.org

COMMITTEE ON CONVENTIONAL PROMPT GLOBAL STRIKE CAPABILITY

ALBERT CARNESALE, University of California, Los Angeles, *Chair*
PAUL J. BRACKEN, Yale University
PAUL K. DAVIS, RAND Corporation
STEVE FETTER, University of Maryland, College Park
JOHN S. FOSTER, JR., Rancho Palos Verdes, California
EUGENE FOX, USA (retired), McLean, Virginia
ALEC D. GALLIMORE, University of Michigan
RICHARD L. GARWIN, IBM Thomas J. Watson Research Center
EUGENE E. HABIGER, USAF (retired), University of Georgia
DAVID V. KALBAUGH, Centreville, Maryland
L. DAVID MONTAGUE, Menlo Park, California
ROBERT B. OAKLEY, National Defense University
WALTER B. SLOCOMBE, Caplin & Drysdale
WILLIAM D. SMITH, USN (retired), Independent Consultant
JOHN P. STENBIT, Oakton, Virginia
DAVID M. VAN WIE, Johns Hopkins University, Applied Physics Laboratory
ROBERT H. WERTHEIM, USN (retired), San Diego, California
ELLEN D. WILLIAMS, University of Maryland, College Park

Staff

CHARLES F. DRAPER, Study Director and Director, Naval Studies Board
RAYMOND S. WIDMAYER, Senior Program Officer (as of July 10, 2007)
BILLY M. WILLIAMS, Senior Program Officer (as of August 18, 2007)
IAN M. CAMERON, Associate Program Officer (through May 21, 2007)
MARTA V. HERNANDEZ, Associate Program Officer (as of March 15, 2008)
SUSAN G. CAMPBELL, Administrative Coordinator
MARY G. GORDON, Information Officer
SEKOU O. JACKSON, Senior Project Assistant
SIDNEY G. REED, JR., Consultant

NAVAL STUDIES BOARD

MIRIAM E. JOHN, Livermore, California, *Chair*
DAVID A. WHELAN, The Boeing Company, *Vice Chair*
LEE M. HAMMARSTROM, Applied Research Laboratory, Pennsylvania State University
JAMES L. HERDT, Chelsea, Alabama
KERRIE L. HOLLEY, IBM Global Services
BARRY M. HOROWITZ, University of Virginia
JAMES D. HULL, Annapolis, Maryland
JOHN W. HUTCHINSON, Harvard University
LEON A. JOHNSON, United Parcel Service
EDWARD H. KAPLAN, Yale University
CATHERINE M. KELLEHER, University of Maryland and Brown University
JERRY A. KRILL, Johns Hopkins University, Applied Physics Laboratory
THOMAS V. McNAMARA, Textron Systems
L. DAVID MONTAGUE, Menlo Park, California
JOHN H. MOXLEY III, Solvang, California
HEIDI C. PERRY, Charles Stark Draper Laboratory, Inc.
GENE H. PORTER, Nashua, New Hampshire
JOHN S. QUILTY, Oakton, Virginia
J. PAUL REASON, Washington, D.C.
JOHN E. RHODES, Balboa, California
JOHN P. STENBIT, Oakton, Virginia
JAMES WARD, Lincoln Laboratory, Massachusetts Institute of Technology
CINDY WILLIAMS, Massachusetts Institute of Technology
ELIHU ZIMET, Gaithersburg, Maryland

Navy Liaison Representatives

RADM DAN W. DAVENPORT, USN, Office of the Chief of Naval Operations, N81 (through July 25, 2007)
RADM WILLIAM R. BURKE, USN, Office of the Chief of Naval Operations, N81 (as of September 26, 2007)
RADM(S) BRIAN C. PRINDLE, USN, Office of the Chief of Naval Operations, N81 (as of August 25, 2008)
RADM WILLIAM E. LANDAY III, USN, Office of the Chief of Naval Operations, N091 (through August 15, 2008)

Marine Corps Liaison Representative

LTGEN JAMES F. AMOS, USMC, Commanding General, Marine Corps Combat Development Command (through July 2, 2008)
LTGEN GEORGE J. FLYNN, USMC, Commanding General, Marine Corps Combat Development Command (as of July 28, 2008)

Staff

CHARLES F. DRAPER, Director, Naval Studies Board
ARUL MOZHI, Senior Program Officer
RAYMOND S. WIDMAYER, Senior Program Officer (as of July 10, 2007)
BILLY M. WILLIAMS, Senior Program Officer (as of August 18, 2007)
IAN M. CAMERON, Associate Program Officer (through May 21, 2007)
MARTA V. HERNANDEZ, Associate Program Officer (as of March 15, 2008)
SUSAN G. CAMPBELL, Administrative Coordinator
MARY G. GORDON, Information Officer
SEKOU O. JACKSON, Senior Program Assistant

DIVISION ON ENGINEERING AND PHYSICAL SCIENCES

Preface

The Department of Defense's (DOD's) 2007 budget request included $127 million for the Conventional Trident Modification (CTM) program—specifically, for the conversion of two Trident II (D5) missiles on each of the U.S. Navy's 12 deployed nuclear-powered ballistic missile submarines (SSBNs) from nuclear-armed to conventionally armed, to provide a conventional prompt global strike (CPGS) capability.[1,2] For the purposes of this report, "conventional" is defined as non-nuclear, "prompt" is defined as striking within 1 hour of launch, and "global strike" is defined as the ability to strike anywhere in the world within meters of the target. The United States does not currently have military capabilities consistent with these definitions.

The 109th Congress rejected most of the DOD's 2007 budget request for CTM because of concerns regarding "nuclear ambiguity" associated with CTM (i.e., the risk that an observed launch of a conventionally armed missile might be mistaken for the launch of a nuclear-armed missile), as well as a belief that other CPGS systems might better address some of the military, political, and technical issues surrounding CTM.[3] The conference report accompanying Department

[1]See Department of Defense, 2007, *Department of Defense Appropriations Bill,* S. Rept. No. 109-292; and Global Security, 2008, *Conventional TRIDENT Modification (CTM),* Alexandria, Va., April 17.

[2]Acronyms and abbreviations are provided in Appendix A.

[3]Furthermore, in a letter dated February 16, 2007, to Dr. Ralph J. Cicerone, President of the National Academy of Sciences, Senators Daniel K. Inouye and Ted Stevens, Chairman and Ranking Member, respectively, of the Senate Committee on Appropriations, Subcommittee on Defense, stated that "there was widespread, but not universal, agreement [in the Senate] that the Congress should not proceed with the conventional Trident program [and that] critical to the opposition was a belief that

of Defense Appropriations Act, 2007 (Public Law 109-289), requested that the National Academy of Sciences conduct a study "to analyze the mission requirement and, where appropriate, consider and recommend alternatives that meet the prompt global strike mission in the near term (1-2 years), the mid-term (3-5 years), and the long term. The study should include analyses of the military, political and international issues associated with each alternative. The study should consider technology options for achieving desired objectives as well as mitigating policy concerns."[4]

After further discussions with congressional staff and DOD officials regarding the origins, scope, timing, and deliverables associated with this congressionally mandated study, the National Research Council (NRC), under the auspices of the Naval Studies Board, established the Committee on Conventional Prompt Global Strike Capability[5] in early 2007 to conduct the study.[6] This report constitutes the committee's final report.

After initial data gathering and deliberations, appropriate NRC review, and DOD security review, the committee provided to the Congress in May 2007 its interim letter report, including among its recommendations:

> In FY 08, fund research, development, testing, and evaluation (RDT&E) efforts associated with CTM at a level sufficient to determine its effectiveness, but in FY 08 withhold funding for full-scale production and deployment (except any that is necessary for testing).[7]

Subsequently, the committee chair, at the invitation of the House Armed Services Committee, met with congressional representatives in July 2007 to discuss the issues raised in the interim letter report, including the issue of nuclear ambiguity, as well as the letter report's key findings and recommendations.

At the time of the initial drafting of this final report (September 2007), the committee faced uncertainty as to the direction that the Congress would take in 2008 on CPGS, and even more uncertainty as to the levels of funding that the Congress would appropriate to CTM and other specified alternative CPGS systems and to DOD-wide efforts on CPGS in general. As it turned out, the conference report accompanying Department of Defense Appropriations Act, 2008 (Public Law 110-116), provided no funding for testing, fabrication, or deployment of CTM,

---

the Trident option proposed the most difficult challenges of ambiguity." This letter is reproduced in Appendix B.

[4]*Making Appropriations for the Department of Defense for the Fiscal Year Ending September 30, 2007, and for Other Purposes: Conference Report to Accompany H.R. 5631*, H. Rept. 109-676, pp. 227-228, 109th Cong., 2d sess. (September 25, 2006).

[5]Committee members' biographies are provided in Appendix C.

[6]The statement of task for this study and the congressional language requesting the study are provided in Appendix D.

[7]The committee's interim letter report is reprinted in Appendix E.

choosing instead to provide $100 million "in a new Prompt Global Strike program element within the Research, Development, Test and Evaluation, Defense-Wide appropriation only for development of promising conventional prompt global strike technologies" to be managed by the Under Secretary of Defense for Acquisition, Technology and Logistics.[8]

For reasons elaborated in this report (and in its Summary), the committee disagrees with the congressional decision not to fund testing of CTM in 2008, and recommends instead that Congress fund CTM RDT&E at a level sufficient to achieve early deployment if tests confirm system effectiveness.

Putting aside these congressional issues and the timing of this study, the committee believes that it has responded productively to the original congressional tasking by providing in this final report a comprehensive analysis of the military, political, and technical issues surrounding the CTM program as well as a thorough analysis of alternative CPGS systems. The committee thanks the many briefers who presented information essential to the writing of this report.[9] In particular, the committee is especially grateful to Gregory Hulcher, Director for Special Projects, Office of the Under Secretary of Defense (Acquisition, Technology and Logistics), and James Colasacco, Division Chief, Global Strike Capability, U.S. Strategic Command—both of whom facilitated the committee's efforts in gathering DOD information related to the study tasking, such as military and technical information related to CTM and CPGS alternatives.

Finally, a supplement to this report, which does not modify any of the report's findings, conclusions, or recommendations, has been produced; the supplement may contain information that the U.S. government and the National Academy of Sciences have determined is not releasable to the public because it would disclose matters described in 5 U.S.C. Section 552(b). Requests for copies of the supplement, which will be considered on a case-by-case basis, may be addressed to the National Research Council's Naval Studies Board (http://www.nas.edu/nsb).

---

[8] *Making Appropriations for the Department of Defense for the Fiscal Year Ending September 30, 2008, and for Other Purposes: Conference Report to Accompany H.R. 3222,* H. Rept. 110-434, p. 240, 110th Cong., 1st sess. (November 6, 2007).

[9] During the course of its study, the committee held meetings in which it received (and discussed) materials that are exempt from release under 5 U.S.C. 552(b). A summary of the committee's meeting agendas is provided in Appendix F.

Acknowledgment of Reviewers

This report has been reviewed in draft form by individuals chosen for their diverse perspectives and technical expertise, in accordance with procedures approved by the National Research Council's Report Review Committee. The purpose of this independent review is to provide candid and critical comments that will assist the institution in making its published report as sound as possible and to ensure that the report meets institutional standards for objectivity, evidence, and responsiveness to the study charge. The review comments and draft manuscript remain confidential to protect the integrity of the deliberative process. We wish to thank the following individuals for their review of this report:

John F. Ahearne, Sigma Xi, The Scientific Research Society,
Edward G. Anderson III, LTG, USA (retired), Booz Allen Hamilton,
Paul M. Bevilaqua, Lockheed Martin Aeronautics Company,
Gregory S. Martin, Gen, USAF (retired), Arlington, Virginia,
Richard W. Mies, ADM, USN (retired), Fairfax Station, Virginia,
Neil G. Siegel, Northrop Grumman Mission Systems, and
Larry D. Welch, Gen, USAF (retired), Alexandria, Virginia.

Although the reviewers listed above provided many constructive comments and suggestions, they were not asked to endorse the conclusions or recommendations, nor did they see the final draft of the report before its release. The review of this report was overseen by Stephen Berry of the University of Chicago. Appointed by the National Research Council, he was responsible for making certain that an independent examination of this report was carried out in accordance with institutional procedures and that all review comments were carefully considered. Responsibility for the final content of this report rests entirely with the authoring committee and the institution.

Contents

APPENDIXES

Summary

The National Research Council's (NRC's) Committee on Conventional Prompt Global Strike Capability[1] was established in response to the conference report accompanying Department of Defense Appropriations Act, 2007 (Public Law 109-289).[2] In May 2007, the committee provided an interim letter report to the Congress which, together with this final report, satisfies the original congressional tasking.[3]

The committee first convened in February 2007 and met over a period of 8 months.[4] In total, the committee received nearly 100 documents from a wide range of organizations, including the Defense Intelligence Agency, Department of State, Joint Chiefs of Staff, Office of the Secretary of Defense, U.S. Air Force, U.S. Army, U.S. Navy, U.S. Strategic Command, and a number of nongovernmental organizations. In addition to its data gathering, the committee met in closed session throughout this study in order to deliberate on its findings and recommendations and to prepare both its interim and final reports. As with the interim report, all findings and recommendations in this final report are supported unanimously by the members of the committee.

---

[1]Committee members' biographies are provided in Appendix C.

[2]*Making Appropriations for the Department of Defense for the Fiscal Year Ending September 30, 2007, and for Other Purposes: Conference Report to Accompany H.R. 5631,* H. Rept. 109-676, pp. 227-228, 109th Cong., 2d sess. (September 25, 2006).

[3]The statement of task for this study and the congressional language requesting the study are provided in Appendix D. The interim letter report is reprinted in Appendix E.

[4]During the course of its study, the committee held meetings in which it received (and discussed) materials that are exempt from release under 5 U.S.C. 552(b). A summary of the committee's meeting agendas is provided in Appendix F.

To frame the analysis of conventional prompt global strike (CPGS), the committee posed the following questions:

1. Does the United States need CPGS capabilities?
2. What are the alternative CPGS systems, and how effective are they likely to be if proposed capabilities are achieved?
3. What would be the implications of alternative CPGS systems for stability, doctrine, decision making, and operations?
4. What nuclear ambiguity concerns arise from CPGS, and how might they be mitigated?
5. What arms control issues arise with CPGS systems, and how might they be resolved?
6. Should the United States proceed with research, development, testing, and evaluation (RDT&E) of the Conventional Trident Modification (CTM) program[5] and, ultimately, with CTM production and deployment?
7. Should the United States proceed with the development and testing of alternative CPGS systems beyond CTM?

Issues surrounding these key questions are discussed in detail throughout the report. To help frame the committee's major recommendations, these questions, along with major findings, are summarized in the following section.

KEY QUESTIONS AND MAJOR FINDINGS RELATING TO CPGS

1. Does the United States Need CPGS Capabilities?

The committee developed a set of credible scenarios and cases with which to assess the feasibility and value of various levels of coverage and promptness and to assess the relative merits of alternative approaches to CPGS. In doing so, it drew on material provided by Department of Defense (DOD) officials, historical experience over the past decade with actual or seriously contemplated strikes, and intelligence projections. The scenarios included, for example, the need to strike a ballistic missile launcher poised to launch a nuclear weapon at the United States or at an ally; an opportunity to strike a gathering of terrorist leaders or a shipment of weapons of mass destruction (WMD) during a brief period of vulnerability; and the need to disable an adversary's command-and-control capability as the leading edge of a broader combat operation.

With the benefit of these scenarios and more specifically defined test cases, the committee concluded that a high-confidence CPGS capability would be valu-

[5]The Conventional Trident Modification program involves the conversion of two Trident II (D5) missiles on each of the U.S. Navy's 12 deployed nuclear-powered ballistic missile submarines from nuclear-armed to conventionally armed.

able; that technical development and assessment should be pursued immediately; and that if system effectiveness is demonstrated, production and deployment should follow as soon as practicable. The committee concluded also that if the DOD's stated goal of achieving "global" strike were to be accepted as a strict criterion, it would rule out potentially attractive options. Long range is an important element of CPGS but not the only factor of interest. Thus, the committee did not interpret the term "global" literally. In contrast, the committee concluded that setting a goal of 1 hour for execution time[6] in a conventional strike was sensible when viewed in terms of feasibility, value, and affordability. But here, too, the goal was not considered as a strict criterion, and some options that would not quite meet the DOD goal were considered in the analysis.

The desire for a CPGS capability has been noted in numerous national defense strategy documents and reports to Congress over the past several years. In February 2007, the Secretary of Defense and Secretary of State submitted to Congress a report that clearly articulated CPGS mission types and made clear the shared DOD and State Department view that CTM is a needed near-term CPGS capability.[7]

At present, U.S. strikes with conventional weapons are conducted primarily through the use of forward-based systems, particularly tactical aircraft and cruise missiles, and with heavy bombers. Effective use of these systems requires that there be adequate time available to position the aircraft and/or missiles within range of the targets, to conduct detailed mission planning, and, when needed, to provide tanker refueling capability. For distances of about 500 nautical miles (nmi) or more, the flight time alone for current air-breathing vehicles exceeds 1 hour. Accordingly, current forward-based systems can meet the "within 1 hour" criterion for a "prompt" strike only for relatively short distances to targets, and then only if appropriately pre-positioned and with extensive mission-support assets available. The growth of sophisticated air defenses might also present problems for forward-deployed forces, unless attacks by those forces were preceded by effective defense-suppression attacks. Figure S-1 displays the CPGS "capability gap" that exists at present and indicates that ballistic missiles, either intercontinental or sea-launched, could potentially fill the gap in terms of the time/distance specifications.

Recent U.S. strikes with conventional weapons have included the use of armed unmanned aerial vehicles (UAVs), such as Predator, in attacks against al-Qaeda members in Pakistan. Clearly, if (1) U.S. forces equipped with armed UAVs are deployed sufficiently close to the targets to enable UAVs to reach the targets "promptly" and if (2) local air defenses do not pose an unacceptable threat to the

[6]"Execution time" is the time between the President's order to execute the attack and when the target is affected.

[7]Secretary of Defense and Secretary of State. 2007. *Report to Congress on Conventional Trident Modification (CTM)* (U), Washington, D.C., February 1 (classified).

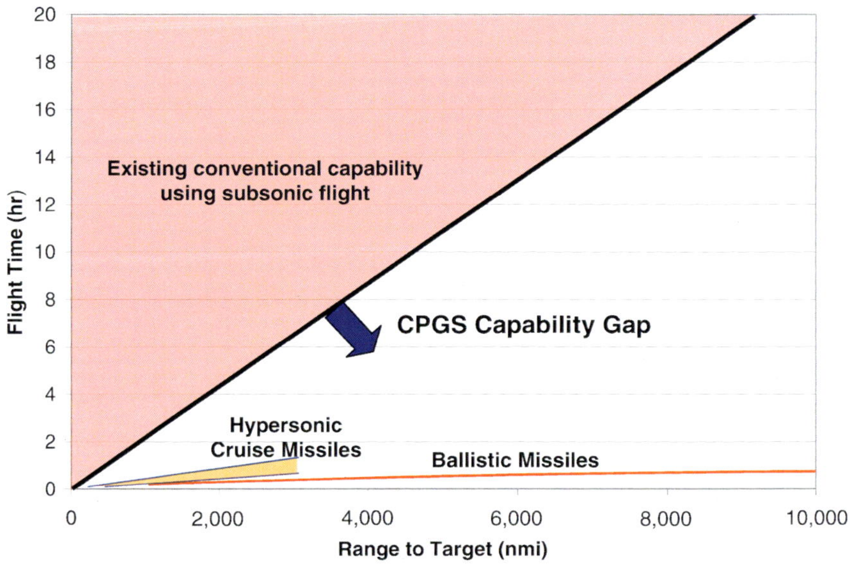

FIGURE S-1 Conventional prompt global strike (CPGS) capability gap. The time needed to reach a target as a function of range for existing conventional systems contrasted with ballistic missiles and hypersonic cruise missiles.

success of the mission, then armed UAVs would provide an option for prompt strike. However, there are many credible scenarios in which these conditions are not met. In addition, CPGS systems employing long-range ballistic missiles with high payloads are projected to be effective against a broader class of targets than could be effectively attacked by armed UAVs.

Major Finding 1. There are credible scenarios in which the United States could gain meaningful political and strategic advantages by being able to strike with conventional weapons important targets that could not be attacked rapidly by currently deployed military assets. In light of the appropriately extreme reluctance to use nuclear weapons, conventional prompt global strike (CPGS) could be of particular value in some important scenarios in that it would eliminate the dilemma of having to choose between responding to a sudden threat either by using nuclear weapons or by not responding at all.

2. What Are the Alternative CPGS Systems, and How Effective Are They Likely to Be If Proposed Capabilities Are Achieved?

The U.S. Air Force Space Command serves as the executive agent for the DOD's analysis of alternatives (AoA) for conventional strike and prompt global strike systems. Expected to be completed in May 2008, the DOD AoA examines systems that might be available in the near term, mid-term, and long term: these consist of CTM, with a projected fiscal year (FY) 2010 initial operational capability (IOC)[8] (based on FY 2008 funding); a Conventional Strike Missile (CSM) based in the continental United States (CONUS), with a projected FY 2014/2015 IOC (based on a funded FY 2008 demonstration program); and four conceptual alternatives with projected FY 2020 IOCs. Informed by the DOD's ongoing AoA, the committee elected to analyze seven CPGS system options.[9]

1. *Existing systems.* These comprise tactical aircraft, cruise missiles, other armed unmanned aerial vehicles, and heavy bombers.

2. *Conventional Trident Modification.* CTM is a near-term alternative proposed by the DOD. It involves the conversion of two Trident II (D5) missiles on each of the U.S. Navy's 12 deployed nuclear-powered ballistic missile submarines (SSBNs) from nuclear-armed to conventionally armed. Each converted Trident missile would carry up to four reentry vehicles equipped with advanced navigation, guidance, and control capabilities. Each reentry vehicle would carry a warhead consisting of dispersible kinetic energy projectiles (KEPs).

3. *CTM-2.* CTM-2 is a mid-term alternative conceived by the committee. It would employ conventional warheads on a missile consisting of the first two stages of the three-stage Trident missile. Removal of the third-stage motor would result in increased payload volume and payload options, while still achieving range on the order of 1,000 nmi.

4. *Submarine-Launched Global Strike Missile (SLGSM).* This mid-term to long-term alternative, conceived by the Navy, would build on CTM experience and could carry either multiple KEP warheads or a single, heavier warhead suitable

[8]"Initial operational capability" is defined in *The Department of Defense Dictionary of Military and Associated Terms* as "the first attainment of the capability to employ effectively a weapon, item of equipment, or system of approved specific characteristics that is manned or operated by an adequately trained, equipped, and supported military unit or force" (Department of Defense, Joint Publication 1-02, April 12, 2001, as amended through October 17, 2007).

[9]Similar to the committee's interim letter report, the final report is based on the committee's collective knowledge as well as on input from other experts, both internal and external to the DOD. Appendix C provides biographical information on the committee members, among whom are technical experts familiar with research, development, and acquisition areas related to strategic strike systems. Accordingly, the committee thought it appropriate in some cases, such as in the case of CTM and CSM, to modify currently proposed DOD CPGS systems in ways that would enhance these systems while at the same time placing their projected capabilities on more realistic paths for potential near-, mid-, and/or long-term utility.

for attack against some hard targets. SLGSMs would be launched from existing nuclear-powered guided missile submarines (SSGNs).

5. *Boost-glide missile (initial version, CSM-1).* CSM-1 is a mid-term alternative and committee modification of the Air Force-proposed Conventional Strike Missile concept, which would be launched using a Minotaur ballistic missile (i.e., a modified version of the no longer operationally deployed Peacekeeper). The CSM-1 system would have extended range and substantial maneuvering capability, which would be useful in avoiding undesired overflights of other countries. An initial deployment of CSM-1 would have specified capabilities of 800 seconds of glide time and of delivering either a KEP or a penetrator warhead suitable for use against some hard targets.[10]

6. *Boost-glide missile (second version, CSM-2).* CSM-2 is a long-term alternative and committee modification of the Air Force-proposed CONUS-based missile concept. The specifications for this longer-term, advanced variant of the boost-glide missile would include a 3,000-second glide time to give it a longer range than that of CSM-1, and a capability to slow to speeds appropriate to dispensing multiple munitions and, possibly, to dispensing modules for intelligence, surveillance, and reconnaissance (ISR); battle damage assessment (BDA); and reattack.

7. *Hypersonic cruise missiles.* Hypersonic cruise missiles are long-term alternatives, similar to the AoA's Mach 6 missile, that could be forward-deployed (e.g., on SSGNs or on land) or launched from long-range aircraft. While hypersonic cruise missiles would have only medium range, their specifications would call for considerable capability for terminal-phase dispensing of smart munitions and ISR modules.

Table S-1 summarizes some of the key attributes of each of these seven alternatives, all of which are analyzed and evaluated in this report. The criteria for evaluation include time to implementation, anticipated cost, delivery accuracy, weapons effectiveness, technical risk, proposed performance in various military scenarios, and contribution to the evolution of long-term CPGS capability.

CTM would be effective against some targets of political and military significance that cannot be struck promptly with existing systems. All alternative CPGS systems, which would to varying degrees overcome the limitations of CTM (e.g., inability to destroy hard or buried targets, limited numbers), would reach IOC later

[10]As discussed in Chapter 4, the time line for the CSM effort planned and funded by the U.S. Air Force is optimistic for a program intended to result in a highly reliable, highly effective presidential-release weapon. In the committee's judgment, a prudently scheduled, well-funded program with adequate testing would have an IOC of about 2017 for an initial version of CSM (which the committee refers to as CSM-1) that has an 800-second glide phase and is capable of delivering either KEP or penetrator warheads, and an IOC of about 2022 for a second version (which the committee refers to as CSM-2) that is proposed to have a 3,000-second glide phase and to be capable of dispensing a wide variety of air-launched weapons at high speed or delivery of KEP or penetrator warheads.

than CTM, as shown in Table S-1.[11] CTM-2 could be effective against some hard targets (because its payload volume is greater than that of CTM), but the IOC of CTM-2 would lag that of CTM by 1 or 2 years. SLGSM would provide greater payload than CTM and would offer more flexible terminal trajectory together with more firepower (number of tubes and number of missiles per tube) than CTM, but it would be more costly than CTM or CTM-2 and is projected to lag CTM by 3 to 4 years. Boost-glide missiles—such as the CSM-1 and CSM-2—and hypersonic cruise missiles would provide maneuvering capability for larger payloads, but they would have much higher associated costs and developmental uncertainties and would take at least 8 years from now to bring to IOC.

Major Finding 2. Conventional Trident Modification (CTM) has advantages over alternative CPGS systems in its near-term availability, low development cost, low opportunity cost, low technical risk, and minimal required changes in declared policy or doctrine. While CTM has limitations compared with other CPGS alternatives, it would be effective against many targets that current systems could not engage quickly enough, and it is the only CPGS system that could be available in the near term.

3. What Would Be the Implications of Alternative CPGS Systems for Stability, Doctrine, Decision Making, and Operations?

The essential policy judgment that must be made in selecting any CPGS system and the doctrine for its use is whether the advantages of the new system would outweigh the disadvantages that it presented. The judgment will depend in part on the type of CPGS system under consideration. Some of the challenges are smaller for CTM than for other CPGS systems because of the CTM program's limited scale, CTM's relatively low cost and technological risk, and its direct lineage from the mature Trident system. However, some of the more advanced CPGS systems (if they prove to be technically feasible) could be designed and deployed so as to reduce ambiguity concerns that arise from the use of delivery vehicles and platforms previously associated with nuclear weapons (see the subsection below).

With the introduction of any significant new military capability, the doctrine for use must be clearly defined by policy makers and clearly understood by military commanders and planners. The use of long-range missiles to deliver conventional weapons promptly and accurately enough to damage meaningful targets requires enabling information (e.g., intelligence, command-and-control,

[11] A 2005 National Research Council report entitled *Effects of Nuclear Earth-Penetrator and Other Weapons* concluded, among other things, that many strategic hard and deeply buried targets can only be attacked directly with nuclear weapons. See National Research Council, 2005, *Effects of Nuclear Earth-Penetrator and Other Weapons,* The National Academies Press, Washington, D.C.

TABLE S-1 Summary of Conventional Prompt Global Strike Alternatives Reviewed by the Committee

Alternatives	Origins	Launch Vehicles	Range (Payload-Dependent)[a]	Munitions Payload Capacity[b]	Earliest IOC[c]	20-Year Cost (relative to CTM: billions of 2009 dollars)[d]
Existing systems	USA, USAF, USMC, USN	Cruise missiles, tactical aircraft, and heavy bombers	1,500 to >6,000 nmi	1,000-2,000 lb	Available now	Not applicable
CTM	USN (sea-based)	Trident: D5 (3-stage)	>4,000 nmi	>1,000 lb	2011	1
CTM-2	Committee (sea-based)	Trident: 2-stage	>4,000 nmi[e]	2,000 lb[e]	2013	3
SLGSM	USN (sea-based)	2-stage rocket booster	3,000 nmi	2,000 lb	2014-2015	5-10
Boost-glide missile (CSM-1)	Committee/ USAF (land-based)[f]	Minotaur III	>6,000 nmi	2,000 lb	2016-2020	10-20

Boost-glide missile (CSM-2)	Committee/ USAF (land-based)[f]	Minotaur III	>6,000 nmi (plus additional glide range vs. CSM-1)	2,000 lb	2018-2024	10-25
Hypersonic cruise missiles	USN (sea-based) or USAF (land-based or B-52)	Single-stage rocket booster	2,000-3,000 nmi	1,000-2,000 lb	2020- 2024	10-20

NOTE: Acronyms are defined in Appendix A.

[a]Data on range and payload for CTM, SLGSM, CSM-1, CSM-2, and hypersonic cruise missile options are extracted from Amy F. Woolf, 2007, *Conventional Warheads for Long-Range Ballistic Missiles: Background and Issues for Congress,* CRS Report to Congress, Congressional Research Service, Washington, D.C., June 19, pp. 10-12, 24-26.

[b]The reader is cautioned that direct mass-to-mass comparisons of munitions capacity do not reflect weapons effectiveness. Different types of munitions will have different weapon impact for the same mass.

[c]The reported initial operational capability (IOC) data in this table are the committee's best estimates based on information presented to the committee and the experience of committee members, assuming an authorization date of 2008. Actual IOCs for all but the CTM are likely to be later for many reasons, including delays in decision making, the time required to stand up program offices, and unanticipated problems in systems engineering.

[d]The 20-year cost estimates are based on contractor briefings. The numbers quoted are imprecise estimates of costs relative to the projected cost for CTM.

[e]The committee-generated CTM-2 concept would have a larger payload capability due to the throw weight and volume freed up by removing the third-stage motor of a Trident missile. Range, however, would be somewhat lower depending on payload.

[f]CSM-1 and CSM-2 are committee modifications of the Air Force-proposed CSM and CONUS missile concepts, respectively.

targeting) that is significantly more difficult to obtain than that needed for nuclear weapons or for conventional weapons delivered by manned aircraft.[12] The speed of delivery of enabling information and of a decision to execute must be fast enough to achieve an extremely short overall execution time, yet rigorous command-and-control requirements must be met. The committee believes that a fundamental command-and-control imperative for CPGS systems would be that the weapon could be employed only on the order of the President of the United States. A comprehensive study of the military and diplomatic implications of acquiring and possibly employing CPGS capabilities should precede any deployment and should include the consideration of factors such as the potential for inappropriate, mistaken, or accidental use; the implications for nuclear deterrence and crisis stability (including ambiguity considerations); the impact of overflight and debris; and the implications for arms control and associated agreements.

Major Finding 3. CPGS systems raise policy, doctrine, and operational issues that should be studied comprehensively prior to deployment. The committee's examination of these issues leads it to believe that they could be resolved satisfactorily and that they would not be an obstacle to deployment. Areas of comprehensive study should include the potential for inappropriate, mistaken, or accidental use; the implications for nuclear deterrence and crisis stability (including "ambiguity" considerations); the impact of overflight and debris; and the implications for arms control and associated agreements.

4. What Nuclear Ambiguity Concerns Arise from CPGS, and How Might They Be Mitigated?

Nuclear ambiguity is the most frequently raised objection to proceeding with CTM.[13] This concern is often described as arising only in the case of CPGS systems that (like CTM) would use delivery vehicles and platforms previously associated with nuclear weapons. However, the concern applies to varying degrees to any CPGS system, since any vehicle capable of delivering a conventional

[12]Ballistic missile delivery systems were developed for nuclear weapons for which, owing to the large damage area of the weapon, achieving the required accuracy in the placement of the weapon is relatively easy to accomplish. In contrast, for conventional weapons, accuracy of placement (technically referred to in terms of the circular error probable [CEP] or spherical error probable [SEP]) is essential in order to obtain the desired effects on the target.

[13]In a letter dated February 16, 2007, to Dr. Ralph J. Cicerone, President of the National Academy of Sciences, Senators Daniel K. Inouye and Ted Stevens, Chairman and Ranking Member, respectively, of the Senate Committee on Appropriations, Subcommittee on Defense, stated that "there was widespread, but not universal, agreement [in the Senate] that the Congress should not proceed with the conventional Trident program [and that] critical to the opposition was a belief that the Trident option proposed the most difficult challenge of ambiguity." This letter is reproduced in Appendix B.

weapon to long range and with high speed and accuracy would also be effective in delivering a nuclear weapon.

If another country, for example Russia or China, were to detect the launch of one or more conventionally armed long-range missiles from a deployed SSBN, how might it interpret the event?

There are two aspects, logically and practically distinct, of the nuclear ambiguity issue. The first is the possible misinterpretation by an observing nation of a conventional strike on a third party as a nuclear strike on its own territory. The second is the possible misinterpretation by an observing nation of one or more conventionally armed missiles headed toward its territory as a nuclear attack. The ambiguity issue is more significant in the second case. The committee's analysis of the nuclear ambiguity issue focused on the following questions:

- Who would be able to detect the launch?
- If a foreign nation were to detect the launch, would it be able to identify correctly the missile type and estimate the trajectory of the missile or the reentry vehicles?
- If a launch were detected by a foreign nation, what would happen?
- Would that nation's nuclear forces and surveillance systems be alerted, and if so what would be the consequences?
- Would a "retaliatory" strike be ordered?
- Even if there were no immediate adverse effects, what long-term reactions might be triggered?

These questions are stated simply, but the answers do not lend themselves to such simplicity. Nevertheless, the committee's judgment is as follows:

- In the next 5 years or so, only Russia will be able to detect the launch of a ballistic missile from a sea-based platform. China might obtain such capability soon thereafter.
- If a foreign nation were able to observe the launch of a ballistic missile, it might also be able to determine the missile type. If it were to track the early flight of the missile, it would be able to predict the subsequent ballistic trajectory of the reentry vehicles. Predicting the course of a maneuverable system would be far more difficult.
- The reaction of a nation observing the launch would depend on the context. For example, is the observing nation an adversary of the United States or an ally? Does the missile appear to be headed to a target on its territory? Is this event occurring in the midst of a period of conflict or in a time of relative peace?
- Command and surveillance systems would likely be fully alerted and, depending on the context, military forces (possibly including nuclear forces) might also be brought to higher alert status.
- The committee believes that the risk of the observing nation's launching

a nuclear retaliatory attack is very low. A foreign nation would be extremely unlikely to believe that the United States was starting a nuclear war with only a handful of missiles, and that nation would have every incentive, in its own interest, to determine definitively the character of the attack before responding.

- This risk could be reduced even further by means of cooperative measures, such as providing information to bilateral partners about the CPGS system, its operation, and the doctrine for its use; immediately notifying of launches against countries; and installing devices (such as continuous monitoring systems) to increase the confidence that conventional warheads had not been replaced by nuclear warheads.[14]
- The possibility of conditions in which misinterpretations would be plausible is not, in the committee's judgment, a valid reason to forgo the CPGS capability for those many other cases in which the risk of misinterpretation is negligible.
- Substantive (as opposed to rhetorical) international reactions to the U.S. acquisition and possible use of CPGS capabilities probably would include countermeasures intended to protect valuable potential targets and might include increased emphasis on acquiring comparable conventional strike capabilities. The intensity of any such reactions would be expected to depend to a substantial degree on the scale of the U.S. CPGS deployment or use. For example, the detection of the launch of a large number of conventionally armed missiles might be interpreted as an attempted "disarming first strike" against a nation's strategic forces. The committee believes that the CTM program, as currently envisioned, is sufficiently small in scale to make it unlikely that international reactions would be of strategic significance.

Major Finding 4. Nuclear ambiguity is an understandable concern regarding CTM and, to varying degrees, all other CPGS systems. Nuclear ambiguity cannot be eliminated simply by avoiding a "legacy" nuclear system, such as Trident. The risk of a CPGS attack being misinterpreted and leading to a nuclear attack on the United States could be mitigated and managed through readily available mechanisms. The benefits of possessing a limited CPGS capability, such as that provided by CTM, outweigh the risks associated with nuclear ambiguity.

5. What Arms Control Issues Arise with CPGS Systems, and How Might They Be Resolved?

Both the Intermediate-Range Nuclear Forces (INF) Treaty of 1987 and the Strategic Arms Reduction Treaty (START) of 1991 contain provisions that would

[14]Appendix H provides a discussion of some cooperative efforts (both technical and nontechnical) that can be applied toward mitigating nuclear ambiguity.

apply to certain CPGS systems. The Moscow Treaty of 2002 (Strategic Offensive Reductions Treaty, or SORT) requires reduction and limitation of strategic nuclear warheads, but it does not constrain non-nuclear warheads and, therefore, does not affect any CPGS system.

The INF Treaty had been signed in 1987 by the United States and the former Soviet Union; after the breakup of the Soviet Union, Russia, Belarus, Kazakhstan, and Ukraine assumed the obligations of the former Soviet Union. That treaty prohibits flight-testing, production, and deployment of ground-launched ballistic missiles and ground-launched cruise missiles with ranges between 500 and 5,500 kilometers, regardless of warhead type. It places no restrictions on manned aircraft, air-launched or sea-launched systems, or on ground-launched systems with ranges less than 500 or greater than 5,500 kilometers. The INF Treaty is of unlimited duration.

START requires the United States and Russia to limit their deployed strategic arsenals to no more than 6,000 warheads, with no more than 4,900 on intercontinental ballistic missiles (ICBMs) and submarine-launched ballistic missiles (SLBMs), in accordance with agreed counting rules. START covers all ICBMs and SLBMs and their associated launchers, including new types of ballistic missiles, and no distinction is made between nuclear-armed and non-nuclear-armed ballistic missiles. START is scheduled to remain in force until December 5, 2009, but even if not extended or renegotiated, it is unlikely that its restrictions will disappear entirely.

With these arms control issues in mind, an obvious question for some readers may be why some of the Air Force land-based Minuteman III missiles should not be converted from nuclear-armed to conventionally armed in a manner similar to the Navy CTM program. After thorough investigation of the Minuteman option, the committee concluded that, although technically viable, it is not a realistic contender (for reasons described in Appendix I). In short, the required renegotiation of START, combined with "not-in-my-backyard" issues, presents substantial challenges to deploying ICBMs in locations such as Hawaii or Guam, particularly in the near term. These challenges are judged to be significantly more formidable than the obstacles to CTM deployment. Neither minimizing the number of conventionally armed ICBMs to be deployed in Hawaii or Guam nor using mobile units rather than silos altered the committee's conclusion.[15]

Major Finding 5. Neither CTM nor CTM-2 is in conflict with the Intermediate-Range Nuclear Forces (INF) Treaty of 1987 or the Strategic Arms

[15]In the committee's judgment, any proposed land-based, conventional ICBM system, such as one based on the no longer operationally deployed Peacekeeper, would face the same near-term concerns for deployment as those outlined in Appendix I for Minuteman. These near-term concerns include the likely renegotiation of START, thereby raising the possibility of complications, delays, and uncertainties, as well as the political, economic, and procedural difficulties of introducing new and secure operational bases for strategic missiles.

Reduction Treaty (START) of 1991 because (1) the INF Treaty applies only to ground-launched missiles, and (2) even though the START rules for counting compliance with its limits on missiles, launchers, and warheads would apply to CTM and CTM-2, total planned U.S. deployments of systems subject to those limits are sufficiently below START limits to allow for the envisaged deployments of CTM or CTM-2. Other CPGS systems could raise arms control issues if the INF Treaty and START remain in force or are renegotiated without including provisions that would permit the deployment of these systems.

6. Should the United States Proceed with Research, Development, Testing, and Evaluation of the CTM Program and, Ultimately, with CTM Production and Deployment?

As indicated earlier, the committee believes that there exist plausible scenarios in which CTM would be a valuable and reasonably early addition to U.S. military capabilities, and that CTM should be deployed as quickly as possible, provided that the system's planned effectiveness is demonstrated in tests and that international and political concerns are appropriately mitigated and managed. Accordingly, the committee believes that the United States should proceed with CTM RDT&E to demonstrate system effectiveness (especially with respect to achieving accuracy on the order of meters and lethal effects on the classes of targets of interest), while concurrently pursuing cooperative measures by which to address and mitigate the concerns of Russia and other countries. The committee believes also that the proposed RDT&E program for CTM should be expanded to include the consideration of reentry vehicles capable of diving vertically (in order to attack targets located, for example, on the far side of hills or buildings). Full-scale production and deployment of CTM should proceed only after tests have confirmed that the projected effectiveness of the system has been achieved.

As for the potential for meeting challenges facing CTM, the committee believes (1) that deployment of CTM (i.e., a total of 24 Trident II [D5] missiles converted from nuclear-armed to conventionally armed) would not adversely affect the nation's nuclear deterrent for the foreseeable future; (2) that the Navy has proposed reasonable technical and procedural safeguards against accidental or mistaken launch of a nuclear-armed Trident missile from an SSBN loaded with both nuclear-armed and conventionally armed missiles (and that these safeguards should be strengthened even further in accordance with lessons learned from the Air Force incident in 2007 involving the mistaken transport of nuclear-armed cruise missiles on a B-52 bomber,[16] and that these safeguards should be tested to ensure their effectiveness); (3) that because the fall of early rocket stages from

[16]Joby Warrick and Walter Pincus. 2007. "Missteps in the Bunker," *Washington Post*, September 23.

the CTM would be predictable, any issues associated with overflight could be adequately taken into account by decision makers; (4) that CTM would survive and penetrate likely missile defenses for the foreseeable future; and (5) that the nuclear ambiguity problem inherent in CPGS systems can be adequately managed in the case of CTM.

Finally, the committee notes that CTM, in comparison with alternative CPGS systems under consideration by the DOD and Congress, would be available significantly sooner, with far less cost and far less development risk and uncertainty.

Major Finding 6. The military and political issues associated with CTM appear to be manageable. The CTM research, development, testing, and evaluation (RDT&E) program would be useful not only in demonstrating the effectiveness of CTM as a precondition to full-scale production and deployment, but also in providing technology and flight experience for more advanced CPGS systems. Failure to proceed with CTM RDT&E could significantly delay progress on other CPGS systems.

7. Should the United States Proceed with the Development and Testing of Alternative CPGS Systems Beyond CTM?

CPGS options employing SLBMs are attractive in the near term (CTM), mid-term (CTM-2), and long term (SLGSM) and offer evolutionary paths that balance technical risk with the rapid fielding of improved capabilities.[17] In the long term, boost-glide missiles and hypersonic cruise missiles offer versatile capabilities, but they have high cost and high technical risk and may be less able than ballistic missiles are to penetrate defenses. Land-based ballistic missile options offer little that SLBMs do not, but their higher payloads would enable them to carry conventional warheads better suited to attacking some hard targets, albeit with larger technical risk in obtaining the necessary targeting accuracy.

The committee believes that the United States should proceed with technology development to support the mid- and long-term CPGS options; however, any decisions to proceed with full-scale testing or beyond should be made in the

[17]Whether a growth path from CTM or a first-step development itself, the CTM-2 concept would be a significant advance beyond CTM because it would have greater payload capacity. It seems possible that CTM-2 could be available in 5 to 6 years from program start, although a full engineering analysis of the option would be needed to confirm the timescale. CTM-2's two-stage propulsion could distinguish it from the three-stage Trident, but with high confidence today probably only when tracked by the United States' own systems. Some members of the committee felt strongly that if a possible evolutionary path for CTM-2 were taken, the next generation (beyond CTM-2) could incorporate a reentry vehicle that could deliver (yet-to-be-developed) air-breathing vehicles into theater so as to provide capabilities against moving targets and for post-attack assessment and reattack. Such an evolutionary path is an important approach toward balancing technical risk with rapid fielding of improved CPGS capabilities.

broader context of the nation's policy on strategic strike. The policy on strategic strike must be consistent with the U.S. national security strategy, which the next presidential administration will review in 2009. Development and deployment of any CPGS option beyond CTM or CTM-2 would require a very large national investment.

Major Finding 7. CTM-2 is worth exploring as a mid-term successor to CTM. Of the long-term alternative CPGS systems, the Submarine-Launched Global Strike Missile (SLGSM) appears to have the lowest technical risk and to offer important capabilities, such as the ability to launch from existing, dedicated, conventional strike platforms (nuclear-powered guided missile submarines, or SSGNs). Boost-glide missiles and hypersonic cruise missiles have higher technical risk but, if demonstrated, could provide some advantages beyond submarine-launched ballistic missiles (SLBMs) (e.g., in payload, versatility, and maneuverability).

MAJOR RECOMMENDATIONS

Based on its analysis of CPGS, the committee offers the following recommendations.

Near Term (1 to 2 Years)

In the near term (the next 1 to 2 years), the committee recommends that the following be done:

- Fund CTM research, development, testing, and evaluation (RDT&E) at a level sufficient to achieve early deployment if tests confirm system effectiveness. In parallel, develop doctrine and policies for use of the system, including ensuring the availability of the necessary intelligence and other enablers, and for decision-making procedures.
- Fund exploration of the potential of CTM-2 and, if it is deemed promising, fund RDT&E for CTM-2.
- Fund technology development to explore longer-term CPGS weapon delivery options.
- Initiate programs to improve targeting, planning, and decision making to support CPGS capability, and establish decision points about where to go beyond CTM.
- Initiate what will likely be a multiyear study of concepts and doctrine for potential larger-scale CPGS deployments, recognizing that going beyond the niche capability anticipated currently would raise profound issues that have not yet been adequately explored and debated.

Mid-Term (3 to 5 Years)

The committee recommends that for the mid-term (the next 3 to 5 years), the following be carried out:

- If CTM is demonstrated to be effective, particularly its ability to achieve the necessary accuracy and warhead performance, fund full-scale production and deployment of CTM.
- If CTM-2 is demonstrated to be significantly more effective than CTM, fund the production and deployment of CTM-2 as a follow-on to CTM.
- Continue to fund technology development for the more promising longer-term CPGS options.
- Work with allies, Russia, and others to mitigate policy and international concerns associated with a U.S. CPGS capability, including establishing cooperative measures to reduce ambiguity risks.
- Conduct a comprehensive examination of strategic strike (nuclear and conventional)—across a full spectrum of scenarios—after review of the national security strategy by the next administration, and determine the implications for the development and possible deployment of advanced CPGS systems and associated enabling systems.

Long Term (Beyond 5 Years)

The committee recommends that in the long term (beyond 5 years), the following be done:

- Explore a range of CPGS options and fund RDT&E of the most promising options at levels sufficient to improve weapon effectiveness (especially maneuverability, accuracy, range, and lethality).
- Continue to work closely with U.S. military and diplomatic representatives, coalition forces, allies, and others to mitigate policy and international concerns as they may arise.

1

Introduction

The 2006 Quadrennial Defense Review (QDR), conducted internally by the Department of Defense (DOD) to identify military capabilities that would contribute to fulfilling U.S. national security needs, stated the following:

> To help shape the choices of countries at strategic crossroads, strengthen deterrence, and hedge against future strategic uncertainty, the Department will develop a wider range of conventional and non-kinetic deterrent options while maintaining a robust nuclear deterrent. It will convert a small number of Trident submarine-launched ballistic missiles for use in conventional prompt global strike.[1]

The 2006 QDR went on to call for the deployment, within 2 years, of an "initial capability to deliver precision-guided conventional warheads using long-range Trident Submarine-Launched Ballistic Missiles."[2]

The DOD's 2007 budget request included $127 million for the Conventional Trident Modification (CTM) program—specifically, for the conversion of two Trident II (D5) missiles on each of the U.S. Navy's 12 deployed nuclear-powered ballistic missile submarines (SSBNs) from nuclear-armed to conventionally armed, to provide a conventional prompt global strike (CPGS) capability. For the purposes of this report, "conventional" is defined as non-nuclear, "prompt" is defined as striking within 1 hour of launch, and "global strike" is defined as the ability to

[1]Department of Defense. 2006. *Quadrennial Defense Review Report: February 6, 2006,* Washington, D.C., p. 6.

[2]Department of Defense. 2006. *Quadrennial Defense Review Report: February 6, 2006,* Washington, D.C., p. 50.

strike anywhere in the world within meters of the target. The United States does not currently have military capabilities consistent with these definitions.

The 109th Congress rejected most of the DOD's 2007 budget request for CTM because of concerns regarding "nuclear ambiguity" associated with CTM (i.e., the risk that an observed launch of a conventionally armed missile might be mistaken for the launch of a nuclear-armed missile), as well as a belief that other CPGS systems might better address some of the military, political, and technical issues surrounding CTM.[3] The conference report accompanying Department of Defense Appropriations Act, 2007 (Public Law 109-289), requested that the National Academy of Sciences conduct a study "to analyze the mission requirement and, where appropriate, consider and recommend alternatives that meet the prompt global strike mission in the near term (1-2 years), the mid-term (3-5 years), and the long term. The study should include analyses of the military, political and international issues associated with each alternative. The study should consider technology options for achieving desired objectives as well as mitigating policy concerns."[4] In addition, the conference report went on to state that $20 million in research, development, testing, and evaluation (RDT&E)[5] would be provided to the Navy for CTM efforts and that "these funds should be used to focus on those developmental items which are common to all the global strike alternatives until the completion of the study and a determination has been made on the best course of action in this matter."[6]

After further discussions with congressional staff and DOD officials regarding the origins, scope, timing, and deliverables associated with this congressionally mandated study, the National Research Council (NRC), under the auspices of the Naval Studies Board, established the Committee on Conventional Prompt Global Strike Capability in early 2007 to undertake a 15-month study necessary to produce two reports: (1) an interim letter report summarizing the requirements

---

[3]Furthermore, in a letter dated February 16, 2007, to Dr. Ralph J. Cicerone, President of the National Academy of Sciences, Senators Daniel K. Inouye and Ted Stevens, Chairman and Ranking Member, respectively, of the Senate Committee on Appropriations, Subcommittee on Defense, stated that "there was widespread, but not universal, agreement [in the Senate] that the Congress should not proceed with the conventional Trident program [and that] critical to the opposition was a belief that the Trident option proposed the most difficult challenges of ambiguity." This letter is reproduced in Appendix B.

[4]*Making Appropriations for the Department of Defense for the Fiscal Year Ending September 30, 2007, and for Other Purposes: Conference Report to Accompany H.R. 5631,* H. Rept. 109-676, pp. 227-228, 109th Cong., 2d sess. (September 25, 2006).

[5]RDT&E activities include basic research, applied research, advanced technology development, advanced component development and prototypes, system development and demonstration, management support (e.g., test ranges), and operational system development, according to Department of Defense, 2007, Office of the Under Secretary of Defense (Comptroller), *DOD Financial Management Regulation, Volume 2B: Budget Formulation and Presentation,* July.

[6]*Making Appropriations for the Department of Defense for the Fiscal Year Ending September 30, 2007, and for Other Purposes: Conference Report to Accompany H.R. 5631*· H. Rept. 109-676. pp. 227-228, 109th Cong., 2d sess. (September 25, 2006).

and supporting enablers[7] for a CPGS capability, and recommending a near-term option or options to provide this capability; and (2) a comprehensive report that addresses the full terms of reference in the original congressional tasking.

After initial data gathering and deliberations, appropriate NRC review, and DOD security review, the committee provided to the Congress in May 2007 its interim letter report, including among its recommendations:

> In FY 08, fund research, development, testing, and evaluation (RDT&E) efforts associated with CTM at a level sufficient to determine its effectiveness, but in FY 08 withhold funding for full-scale production and deployment (except any that is necessary for testing).[8]

Subsequently, the committee chair, at the invitation of the House Armed Services Committee, met with congressional representatives in July 2007 to discuss the issues raised in the interim letter report, including the issue of nuclear ambiguity, as well as the report's key findings and recommendations. In the months during the writing and following the release of its interim letter report, the committee continued gathering information, and in September 2007 it convened to begin drafting its final report.[9]

At the time of the drafting of this final report, the committee faced uncertainty as to the direction that the Congress would take in 2008 on CPGS, and even more uncertainty as to the levels of funding that the Congress would appropriate for CTM and other specified alternative CPGS systems and for DOD-wide efforts on CPGS in general. As it turned out, the conference report accompanying Department of Defense Appropriations Act, 2008 (Public Law 110-116), provided no funding for testing, fabrication, or deployment of CTM, choosing instead to provide $100 million "in a new Prompt Global Strike program element within the Research, Defense, Test and Evaluation, Defense-Wide appropriation only for development of promising conventional prompt global strike technologies" to be managed by the Under Secretary of Defense for Acquisition, Technology and Logistics.[10]

The committee disagrees with the congressional decision not to fund testing of CTM in 2008. Indeed, the committee recommends in Chapter 6 (along with additional major recommendations) that Congress should: "Fund CTM RDT&E at a level sufficient to achieve early deployment if tests confirm system effec-

---

[7]"Supporting enablers" include intelligence support, mission planning, target development, and decision making.

[8]The committee's interim letter report is reprinted in Appendix E of this final report.

[9]During the course of its study, the committee held meetings in which it received (and discussed) materials that are exempt from release under 5 U.S.C. 552(b). A summary of the committee's meeting agendas is provided in Appendix F.

[10]*Making Appropriations for the Department of Defense for the Fiscal Year Ending September 30, 2008, and for Other Purposes: Conference Report to Accompany H.R. 3222,* H. Rept. 110-434. p. 240, 110th Cong., 1st sess. (November 6, 2007).

tiveness." Among the factors leading to the recommendation are that, compared to other CPGS systems, CTM (or a close variant based on the Trident missile, submarine, and associated infrastructure) would be available significantly sooner, at far less cost and with far less development risk and uncertainty. Moreover, a deferral of CTM RDT&E would adversely affect the schedules and development risks associated with other CPGS systems.

THE CPGS CAPABILITY GAP

The committee developed a set of credible scenarios and cases with which to assess the feasibility and value of various levels of coverage and promptness and to assess the relative merits of alternative approaches to CPGS. In doing so, it drew on material provided by DOD officials, historical experience over the past decade with actual or seriously contemplated strikes, and intelligence projections. The scenarios included, for example, the need to strike a ballistic missile launcher poised to launch a nuclear weapon at the United States or at an ally, an opportunity to strike a gathering of terrorist leaders or a shipment of weapons of mass destruction (WMD) during a brief period of vulnerability, and the need to disable an adversary's command-and-control capability as the leading edge of a broader combat operation.

With the benefit of these scenarios and more specifically defined test cases, the committee concluded that a high-confidence CPGS capability would be valuable, that technical development and assessment should be pursued immediately, and that if system effectiveness is demonstrated, production and deployment should follow as soon as practicable. The committee concluded also that if the DOD's stated goal of achieving "global" strike were to be accepted as a strict criterion, it would rule out potentially attractive options. Long range is an important element of CPGS but not the only factor of interest. Thus, the committee did not interpret the term "global" literally. In contrast, the committee concluded that setting a goal of 1 hour for execution time[11] in a conventional strike was sensible when viewed in terms of feasibility, value, and affordability. But here, too, the goal was not considered as a strict criterion, and some options that would not quite meet the DOD goal were considered in the analysis.

The desire for a CPGS capability has been noted in numerous national defense strategy documents and reports to Congress over the past several years. In February 2007, the Secretary of Defense and Secretary of State submitted to Congress a report that clearly articulated CPGS mission types and made clear the shared DOD and State Department view that CTM is a needed near-term CPGS capability.[12]

---

[11]"Execution time" is the time between the President's order to execute the attack and when the target is affected.

[12]Secretary of Defense and Secretary of State. 2007. *Report to Congress on Conventional Trident Modification (CTM)* (U), Washington, D.C., February 1 (classified).

At present, U.S. strikes with conventional weapons are conducted primarily through the use of forward-based systems, particularly tactical aircraft and cruise missiles, and with heavy bombers. Effective use of these systems requires that there be adequate time available to position the aircraft and/or missiles within range of the targets, to conduct detailed mission planning, and, when needed, to provide tanker refueling capability. For distances of about 500 nautical miles (nmi) or more, the flight time alone for current air-breathing vehicles exceeds 1 hour. Accordingly, current forward-based systems can meet the "within 1 hour" criterion for a "prompt" strike only for relatively short distances to targets, and then only if appropriately pre-positioned and with extensive mission-support assets available.[13] The growth of sophisticated air defenses might also present problems for forward-deployed forces, unless attacks by those forces were preceded by effective high-volume defense-suppression attacks.

Recent U.S. strikes with conventional weapons have included the use of armed unmanned aerial vehicles (UAVs), such as Predator, in attacks against al-Qaeda members in Pakistan. Clearly, if (1) U.S. forces equipped with armed UAVs are deployed close enough to the targets to enable UAVs to reach the targets "promptly" and if (2) local air defenses do not pose an unacceptable threat to the success of the mission, then armed UAVs would provide an option for prompt strike. However, there are many credible scenarios in which these conditions are not met. In addition, CPGS systems employing long-range ballistic missiles with high payloads are projected to be effective against a broader class of targets than could be effectively attacked by armed UAVs.

CONVENTIONAL TRIDENT MODIFICATION AND SOME CPGS ALTERNATIVES

The U.S. Air Force Space Command serves as the executive agent for the DOD's analysis of alternatives (AoA) for conventional strike and prompt global strike systems. Expected to be completed in May 2008, the DOD AoA examines systems that might be available in the near term, mid-term, and long term: these to consist of CTM, with a projected fiscal year (FY) 2010 initial operational capability (IOC)[14] (based on FY 2008 funding); a Conventional Strike Missile (CSM)

[13]The hypothetical operational scenarios outlining the need for a CPGS capability call into question the capability of current command, control, communications, computers, intelligence, surveillance, and reconnaissance (C4ISR) systems and architecture to support a CPGS weapon. Additional discussion of these supporting capabilities and their critical role in enabling a reliable and effective CPGS capability is provided in Chapters 2 and 3 of this report.

[14]"Initial operational capability" is defined in *The Department of Defense Dictionary of Military and Associated Terms* as "the first attainment of the capability to employ effectively a weapon, item of equipment, or system of approved specific characteristics that is manned or operated by an adequately trained, equipped, and supported military unit or force" (Department of Defense, Joint Publication 1-02, April 12, 2001, as amended through October 17, 2007).

based in the continental United States (CONUS), with a projected FY 2014/2015 IOC (based on a funded FY 2008 demonstration program); and four conceptual alternatives with projected FY 2020 IOCs. Informed by the DOD's ongoing AoA, the committee elected to analyze seven CPGS system options, all of which are summarized and analyzed in this report.[15] Table 1-1 summarizes some of the key attributes of each of these seven alternatives.

ORGANIZATION OF THE REPORT

The remaining chapters of this report provide detailed analyses and thorough discussion of the military issues; the political, international, policy, and doctrinal issues; and the technical issues raised by conventional prompt global strike.

Chapter 2 begins by defining CPGS and what that definition entails. It then provides a set of scenario classes identified by the committee, along with a list of attributes and test cases for comparing CPGS options. Chapter 2 elaborates the representative set of seven CPGS alternatives reviewed by the committee, along with corresponding analyses and assessments of those alternatives in the context of the scenario classes provided. It concludes by discussing some of the enablers (e.g., command and control, accuracy, targeting) associated with CPGS, as well as surety and safeguard needs.

Chapter 3 discusses the political, international, policy, and doctrinal issues raised by conventional prompt global strike. Among the topics addressed are command and control for CPGS and the requirement of presidential authorization for its use, the potential for inappropriate or mistaken use of CPGS by the United States, the impact of CPGS on the nation's nuclear deterrent and on stability, and the issue of nuclear ambiguity.

Chapter 4 begins with a discussion of the key technical requirements and needs for CPGS with respect to potential targets. The chapter goes on to address the major technology challenges presented by CTM and by the CPGS alternatives reviewed by the committee: these challenges include thermal protection; accuracy of guidance, navigation, and control; and munitions, sensors, and interrelated needs such as in-flight communications. Chapter 4 concludes with a summary of the technology readiness levels and corresponding time frames for CTM and other CPGS systems.

Chapter 5 provides a synthesis of the assessments of military issues; political,

[15]Similar to the committee's interim letter report, the final report is based on the committee's collective knowledge as well as on input from other experts, both internal and external to the DOD. Appendix C provides biographical information on the committee members, among whom are technical experts familiar with research, development, and acquisition areas related to strategic strike systems. Accordingly, the committee felt it appropriate in some cases, such as in the case of CTM and CSM, to modify of currently proposed DOD CPGS systems in ways that would enhance these systems while at the same time placing their projected capabilities on more realistic paths for potential near-, mid-, and/or long-term utility.

TABLE 1-1 Summary of Conventional Prompt Global Strike Alternatives Reviewed by the Committee

Alternatives	Origins	Launch Vehicles	Range (Payload-Dependent)[a]	Munitions Payload Capacity[b]	Earliest IOC[c]	20-Year Cost (relative to CTM: billions of 2009 dollars)[d]
Existing systems	USA, USAF, USMC, USN	Cruise missiles, tactical aircraft, and heavy bombers	1,500 to >6,000 nmi	1,000-2,000 lb	Available now	Not applicable
CTM	USN (sea-based)	Trident: D5 (3-stage)	>4,000 nmi	>1,000 lb	2011	1
CTM-2	Committee (sea-based)	Trident: 2-stage	>4,000 nmi[e]	2,000 lb[e]	2013	3
SLGSM	USN (sea-based)	2-stage rocket booster	3,000 nmi	2,000 lb	2014-2015	5-10
Boost-glide missile (CSM-1)	Committee/ USAF (land-based)[f]	Minotaur III	>6,000 nmi	2,000 lb	2016-2020	10-20

Boost-glide missile (CSM-2)	Committee/ USAF (land-based)[f]	Minotaur III	>6,000 nmi (plus additional glide range vs. CSM-1)	2,000 lb	2018-2024	10-25
Hypersonic cruise missiles	USN (sea-based) or USAF (land-based or B-52)	Single-stage rocket booster	2,000-3,000 nmi	1,000-2,000 lb	2020- 2024	10-20

NOTE: Acronyms are defined in Appendix A.

[a]Data on range and payload for CTM, SLGSM, CSM-1, CSM-2, and hypersonic cruise missile options are extracted from Amy F. Woolf, 2007, *Conventional Warheads for Long-Range Ballistic Missiles: Background and Issues for Congress,* CRS Report to Congress, Congressional Research Service, Washington, D.C., June 19, pp. 10-12, 24-26.

[b]The reader is cautioned that direct mass-to-mass comparisons of munitions capacity do not reflect weapons effectiveness. Different types of munitions will have different weapon impact for the same mass.

[c]The reported initial operational capability (IOC) data in this table are the committee's best estimates based on information presented to the committee and the experience of committee members, assuming an authorization date of 2008. Actual IOCs for all but the CTM are likely to be later for many reasons, including delays in decision making, the time required to stand up program offices, and unanticipated problems in systems engineering.

[d]The 20-year cost estimates are based on contractor briefings. The numbers quoted are imprecise estimates of costs relative to the projected cost for CTM.

[e]The committee-generated CTM-2 concept would have a larger payload capability due to the throw weight and volume freed up by removing the third-stage motor of a Trident missile. Range, however, would be somewhat lower depending on payload.

[f]CSM-1 and CSM-2 are committee modifications of the Air Force-proposed CSM and CONUS missile concepts, respectively.

international, policy, and doctrinal issues; and technical issues presented in earlier chapters. It also discusses the advantages and disadvantages of each CPGS option.

Chapter 6 addresses seven key questions framing the issues associated with the development and deployment of a CPGS capability. The questions posed by the committee are as follows:

1. Does the United States need CPGS capabilities?
2. What are the alternative CPGS systems, and how effective are they likely to be if proposed capabilities are achieved?
3. What would be the implications of alternative CPGS systems for stability, doctrine, decision making, and operations?
4. What nuclear ambiguity concerns arise from CPGS, and how might they be mitigated?
5. What arms control issues arise with CPGS systems, and how might they be resolved?
6. Should the United States proceed with research, development, testing, and evaluation (RDT&E) of the Conventional Trident Modification (CTM) program and, ultimately, with CTM production and deployment?
7. Should the United States proceed with the development and testing of alternative CPGS systems beyond CTM?

The discussion of these issues in Chapter 6 incorporates the report's major findings and is followed by the committee's recommendations.

Appendix A provides a list of acronyms and abbreviations used throughout the report. Appendix B contains the letter from Senators Inouye and Stevens to Dr. Ralph J. Cicerone. Appendixes C through F, respectively, provide biographies of the members of the committee, its statement of task, a copy of the committee's interim letter report, and a summary of the committee meeting agendas. Additional background information on the calculations on boost-glide systems, discussion of cooperative reduction of nuclear ambiguity, and evaluation of the Minuteman option for CPGS are provided in Appendixes G through I, respectively.

2

Military Issues

DEFINING CONVENTIONAL PROMPT GLOBAL STRIKE

It is well accepted that the United States needs conventional strike capability, that such capability should be global in its reach, and that it should be usable on a timely basis. It is less clear, however, precisely how that should be translated into what the Department of Defense (DOD) calls "requirements." How much reach is enough? How fast is prompt? The DOD has expressed its preferred answers, but the Committee on Conventional Prompt Global Strike Capability was asked for an independent assessment.

How Should "Global" Be Defined? How Much Reach Is Enough?

It can be argued that strike capability could be needed against targets anywhere on the globe. However, even those who favor capabilities-based planning rather than planning tied to specific scenarios will acknowledge that it is unlikely that strikes will be necessary on outposts in Antarctica, Patagonia, or Tasmania, or even that the ability to threaten such strikes will be necessary. Further, if the world situation should change in this regard, there would likely be time for redeployment (over days or weeks). Because of such considerations, the laws of physics, and economic considerations, the committee concluded that it did not wish to limit consideration to systems that could strike any point on the globe within 1 hour, or even a number of hours. It was willing to make trade-offs on the matter. Thus, in what follows, "conventional prompt global strike (CPGS)" implies long-range coverage but not necessarily "global" coverage at any given time.

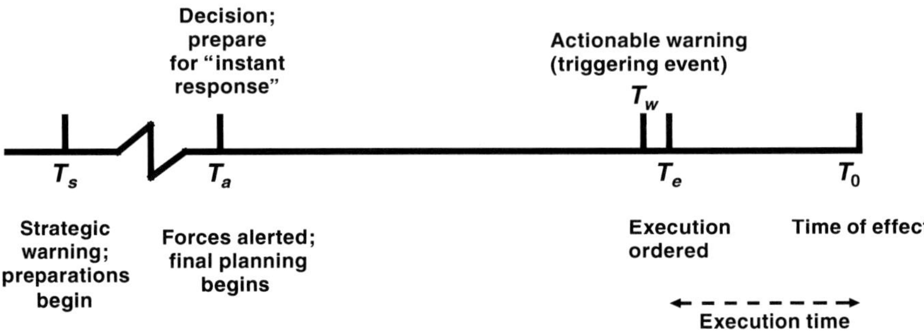

FIGURE 2-1 Time line for a case in which execution follows a substantial period of warning, contingent decision, and preparation, with the final decision made quickly. NOTE: T_s—time of strategic warning; T_a—time at which forces are alerted; T_w—time of actionable warning; T_e—time of execution order; T_0—time of effect. Intelligence and command-and-control activities are critical factors in the overall time line for any CPGS system. The time between actionable warning, T_w, and the order to execute, T_e, includes the time for bringing weapons to ready-to-launch status. SOURCE: Adapted, with permission, from Paul K. Davis, Russell D. Shaver, and Justin Beck, 2008, *Analytical Methods for Assessing Capability Options,* RAND, Santa Monica, Calif., p. 61. © 2008 RAND Corporation.

How Should "Prompt" Be Defined? Is "Within 1 Hour" a Good Criterion?

The same type of question asked about the definition of "global" can be asked about the definition of "prompt." Obviously, an instantaneous strike capability would be desirable if achievable. But what is feasible, valuable, and affordable?

What Is Feasible?

As discussed in detail in Chapter 4 of this report, global strike operations involve a long and complex end-to-end chain of activities. Figure 2-1 explains this chain of activity schematically for a case of particular interest. Two of the most important factors are execution time and the class of targets that can be effectively attacked. "Execution time" is the time between the President's order to execute an attack (at a time T_e) and when the target is affected (at a time T_0).[1] The figure envisions earlier strategic warning (at a time T_s) and an early contingent decision (at a time T_a) that initiates detailed preparations. When "actionable warning" occurs (i.e., when the triggering event occurs, at a time T_w), the final decision to proceed is made quickly (perhaps in tens of minutes, based on precedent, at a time T_e), and execution begins.

[1]This assumes that neither aircraft nor missiles are launched prior to the President's authorization to strike.

Both experience and an examination of possible futures with the DOD's and the committee's scenario-based analysis suggest that this case will be the norm for CPGS: crises and opportunities will seldom arise as complete surprises with targets that merit a conventional "strategic" military strike "popping up." The more important cases are likely to be ones in which events develop over time—perhaps many hours, days, weeks, or even months. As a result, considerable preparation, deliberation, and tentative decision making can occur ahead of time, even if final developments move rapidly—that is, even if execution itself is and needs to be fast (i.e., prompt). In such cases, execution time itself may be the limiting factor *for timeliness.* For example, there could be an advance policy decision to strike at a terrorist leader if he could be located, or at shipments of nuclear materials to a potential proliferator if these materials were identified and located. This said, the other factors are also critical to the mission's success (and will sometimes determine timeliness as well). If CPGS is to be pursued, all aspects of the end-to-end process must be pursued vigorously. Enablers (e.g., intelligence, command and control, and targeting) for CPGS capabilities are discussed later in this chapter. First, however, execution time, target type, and some other key operational factors are addressed.

In addition to execution time, a second key measure is the class of targets against which a CPGS can be used. These classes range from the less difficult (soft point targets such as people or stationary vehicles) to the more difficult (hard and deeply buried target complexes). Other factors (such as whether targets are moving) are also important, but Figure 2-2 uses these first two, execution time and target class, to make some important distinctions.

Figure 2-2 is to be understood as follows. Consider first the solid contour marked "A." This contour indicates that tactical aircraft or cruise missiles are usable for global strike in situations that permit execution times greater than a few hours and which involve targets that may be hard, and may even be buried, but that are not large-area deeply buried targets. This is the region below and to the right of contour A. If the launch platforms are not both located where they need to be *and* poised for action (e.g., if an aircraft carrier is in the appropriate general region but conducting exercises that impede mission execution), the execution time might be more like 10 hours; and if ships or submarines would need to deploy to new locations, the execution time might be days (i.e., more like 100 hours than 10 hours). With existing systems, even the most distant target can be reached in the time it takes for a bomber to fly from Guam, Diego Garcia in the Indian Ocean, or the continental United States (CONUS)—perhaps 10 to 20 hours of actual flight time, plus whatever additional time is needed to generate tanker support.

Moving upward in the figure to the light contour marked "B," note that bombers could have the same range of execution times if they were forward deployed, but if not, execution time would be on the order of 10 to 20 hours (as just described). Bombers, however, would have more capability against complex deeply buried targets because of their great payload.

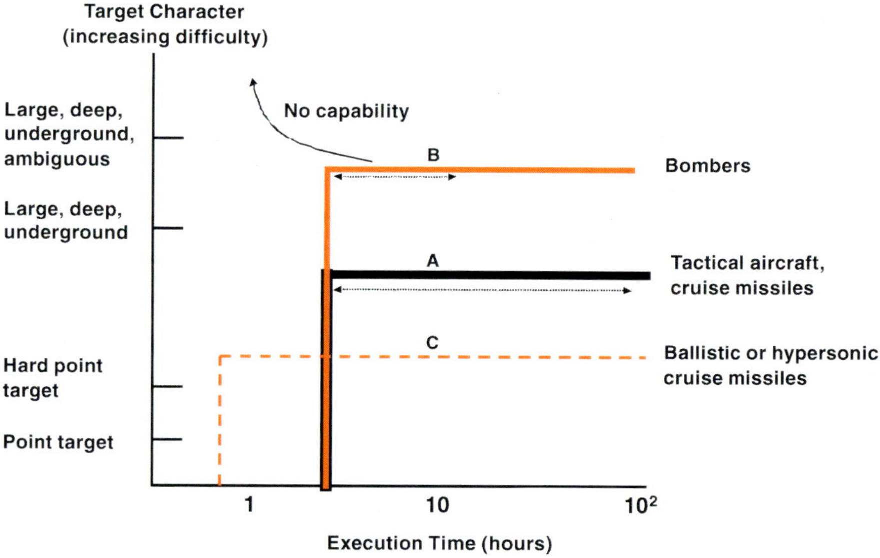

FIGURE 2-2 A depiction of capability space for global strike. NOTE: The dashed lines with two-headed arrows indicate that the boundary moves rightward unless the platforms in question are appropriately deployed and postured. SOURCE: Adapted, with permission, from Paul K. Davis, Russell D. Shaver, and Justin Beck, 2008, *Analytical Methods for Assessing Capability Options,* RAND, Santa Monica, Calif., p. 52. © 2008 RAND Corporation.

Suppose that the mission in question requires shorter execution times. Instantaneous would be ideal, but according to Figure 2-2, the only feasible region is for relatively simple targets and times on the order of 1 hour or more; that is, the dotted contour marked "C." The only weapon systems being contemplated for this region of the capability space for global strike are ballistic missiles and hypersonic cruise missiles.[2]

[2]In special circumstances, any of the contours could move leftward somewhat. If orders were received ahead of time and missiles were at short range, strikes could occur in tens of minutes. If tactical air forces were already in the air and merely had to be diverted to a new target, they could fly 500 miles or so in roughly an hour. For planning purposes, however, more realistic times should be used. Many factors come into play, including maldeployment; being engaged in other missions or not being at a high state of alert; distance from the target; delays associated with receiving, interpreting, and verifying a presidential directive; or the need for detailed at-the-time planning to avoid air defenses, avoid collateral damage, and orchestrate (for some appreciation of such difficulties, see Michael R. Gordon and General Bernard E. Trainor, 2006, *Cobra II: The Inside Story of the Invasion and Occupation of Iraq,* Pantheon Books, New York, Ch. 9). Thus, the committee has treated "a few hours" as a reasonable analytic lower limit for normal air breathers and 1 hour as a reasonable lower limit for ballistic or hypersonic cruise missiles.

What Is Valuable? Would an Execution Time of 1 Hour Be Important?

If a 1-hour execution time is feasible, then the next question is whether it would be of sufficient value to justify the required investment funds and priority. The answer is not immediately obvious; numerous past strikes have been accomplished without such extreme promptness. Nonetheless, the DOD has argued strongly that the capability for prompt strike would be very important, and it has defined "prompt" as an execution time of 1 hour or less. The view of the Secretary of Defense has had firm supportive testimony from the Secretary of State and the Joint Chiefs of Staff. The committee, however, was charged to make an independent assessment, and it concluded that credible instances can readily be imagined for which a CPGS capability would be quite valuable.

Specific scenarios can be useful in judging whether a capability is needed. The committee identified three scenario *classes* (see below), the concreteness of which aided discussion, each of which creates an argument for short execution times such as an hour.[3] These scenario classes are as follows:[4]

1. *Terrorist leaders.* The first scenario class envisions striking a gathering of terrorist leaders expected to be meeting in a particular location for a short period of time. This kind of information is sometimes obtained by electronic intercepts or by intelligence from operatives on the ground. Such a scenario is analogous to actual events from the past decade. In the 1990s, it was sometimes known that al-Qaeda leaders would be meeting, but the time and place were not known precisely until late in the game. In one instance, the President had such intelligence but decided not to strike because of concerns about collateral damage. In another instance, the President authorized a strike in advance, but when the cruise missile subsequently struck the camp, Osama bin Laden was no longer there (having left perhaps several hours earlier according to some reports).[5] During the first phase of the invasion of Iraq in 2003, intelligence reports purportedly located Saddam

[3]Members of the committee differed about which illustrative scenarios should be regarded as the most important and plausible. Some members resonated most with scenario classes 1 and 2, while having deep reservations about class 3. Others resonated with developing deterrence-enhancing capability for class 3 scenarios, while being less persuaded about classes 1 and 2. Overall, the committee believed, of course, that scenarios are merely examples intended to provide concretely some of the many possibilities for which capabilities may be needed. Reasonable people may differ about whether CPGS would be a sensible option to use in any particular scenario, but forgoing the capability would mean not having the option even when it clearly would be needed.

[4]Whether a CPGS capability could actually be employed in the various scenarios would depend on contextual details, such as the strategic environment and political circumstances and, of course, on whether the capabilities ascribed to the CPGS options are successfully achieved and supported by enablers such as described later in this chapter and in Chapter 4.

[5]Many aspects of the event are described in Richard A. Clarke, 2004, *Against All Enemies,* Free Press, New York, Ch. 8. These include the early intelligence, days of preparation and decision making, the detection by Pakistan of some forward-deployed naval forces prior to the attack, and complicated politics.

Hussein at the Iraqi presidential compound at Dora Farms. Aircraft and cruise missiles already in the region were reassigned to strike at the reported site, but Saddam was not there. These historical instances illustrate that execution time can be a critical variable.[6]

2. *WMD shipment.* A second class of scenarios is illustrated by the striking of a transshipment of one or more weapons of mass destruction (WMD), perhaps by truck or ship. Experience tells us that intelligence may exist about when a shipment is planned or may be en route, or where loading, unloading, or temporary stops may occur. Details may be lacking until late—perhaps when those doing the transporting stop for rest or maintenance, or when delays occur at a port, bridge, or border, including stops associated with routine inspections. In some cases, it might be possible for local police or military forces to make an intercept; in other cases, they might not be available to do so, or political considerations might preclude such action. Similarly, political factors might preclude the use of CPGS, as might the risk of collateral damage (caused, for example, by the dissemination of radioactive or noxious materials at the target). In other cases, a prompt strike with U.S. assets might be necessary and appropriate. One such scenario was sketched by former Secretaries of Defense Harold Brown and James Schlesinger in an article supporting CPGS capability.[7]

3. *Immediate response or preemption of imminent attack.* A third class of scenarios is illustrated by imagining that the United States, U.S. forces, or allies are about to be (or have been) attacked and that an immediate response is needed to prevent dire (or additional dire) consequences. Possibilities include conventional or nuclear missile attacks on U.S. cities, on U.S. forces or other military assets, or on U.S. allies. When such scenarios are examined in operational detail, the importance of prompt strike becomes evident: a response time of less than 1 hour might be essential to avert great losses (e.g., loss of numerous satellites crucial to U.S. command and control). In such cases, the CPGS might be the leading edge of a much larger military campaign. That is, the use of CPGS would not have to be large or decisive in itself. It could be very important despite being small. History is again useful. The very earliest part of the military campaign against Iraq in 1991 was a small, specialized, and focused attack on Iraqi air defenses. More generally, military campaigns often begin with relatively small attacks on particular military targets that have leverage—in defeating or disrupting the adversary's air defense; command, control, and communications (C3); or most-feared weapons. And for some of these, timing is critical. With today's forces, such small leading-edge attacks might be conducted by Special Operations Forces (SOF) or other

[6]A sober reading of history also reminds us how frequently human intelligence is wrong.

[7]Harold Brown and James Schlesinger. 2006. "A Missile Strike Option We Need," *Washington Post,* May 22. As an example of how complex a decision to use CPGS could be, consider that an "easy" strike on some platform carrying WMD might not be possible for fear of releasing toxic chemicals or radiation.

forward-deployed assets. In the future, such attacks might be accomplished with prompt global strike assets—with less need for forward basing and less concern about failure due to long flight times or detection.

Is 1-Hour Capability Affordable?

Having established that using a 1-hour criterion for promptness makes sense from the viewpoint of feasibility and that credible cases can be envisioned for which it would be valuable, the remaining issue is cost (and related trade-offs). Without elaboration here, since options on the matter are described later in this chapter and in Chapter 4, it can be stated that achieving 1-hour capability is relatively inexpensive (in DOD terms), although it could be quite expensive if the less-expensive CPGS options are ruled out for one reason or another.

Conclusions on Definitions That Make Sense

In summary, the committee concluded that defining conventional prompt global strike should not be too stringent with respect to the "global" criterion, but that the 1-hour criterion in defining "prompt" was sensible. That said, it could imagine systems (as discussed later) that might not quite make the criterion of promptness but that would be close enough so that they should be considered among the options treated.

ATTRIBUTES AND TEST CASES FOR COMPARING OPTIONS

Attributes

Having discussed execution time and target classes, let us next consider the longer list of factors important to CPGS capability, displayed in Table 2-1. The first three have already been discussed (the second is a subset of the first); the others have not.

Test Cases for Operational Scenarios

The previous section and Table 2-1 outline a way to compare options technically, but a more "operational" comparison is also needed, one more closely related to military utility in actual missions. One way to do this involves constructing a *spanning set* of scenarios construed as test cases. A spanning set needs to stress the alternative systems with respect to timeliness, target class, geographic depth, adversary air defenses, and the need to search and find targets at the time that weapons arrive.

Drawing on the three scenario classes described earlier but thinking about the various operational issues to be considered, the committee constructed a set of six

TABLE 2-1 Variables Important to Conventional Prompt Global Strike

Capability Feature	Meaning
Execution time	Time from the order to execute until effects are achieved (assuming no contingent launch of aircraft or missiles)
Final positioning and planning time	Time from the order to execute until systems are located and supported properly and mission planning is accomplished. May be a part of execution time.
Relevant prompt global strike targets	Target classes (e.g., soft point targets, hard point targets)
Lethality of weapons	The probability of achieving the target damage sought, assuming successful delivery
Ability to attack moving targets	Ability to attack targets that are on the move when attacked
Volume of fire	The number of weapons available for use in a given strike
Controllability	Safety and security with respect to the weapons and their employment
Geographic coverage	The targets that can be reached without repositioning, e.g., without substantial sailing time or rebasing of bombers
Risk of operational side effects	The likelihood or potential for negative effects on, e.g., nuclear deterrent or Special Operations Forces operations
At-the-time strategic factors	Need or non-need for overflight rights, allied cooperation, or forward basing
Availability	Earliest initial operational capability
Development risk	The likelihood of serious development failures or slippages; a function of the technical readiness level of components, full-system testing, industrial base, and organizational competence
Confidence in high reliability	Assuming "successful development," how likely would the system be to have high reliabilities (e.g., 0.95 rather than 0.6)?
Evolutionary potential	Value of the option in laying the technical and operational groundwork for subsequent, more advanced systems

test cases as a reasonably good (but not exhaustive) spanning set. All are variants of the three scenario classes mentioned above, but they have been reformulated to focus on operational issues. They can be summarized as follows:

1a. *Immediate response to threat* (soft targets, near-surface targets, moderate air defenses);

1b. *Immediate initial response to threat* (soft targets, deeply buried targets, advanced air defenses);

2. *Attack of deeply buried WMD facility* (hard, deeply buried targets, advanced air defenses);[8]

3a. *Intercept of terrorism-related WMD transshipment* (soft targets, near-surface targets, moderate or no air defenses, mobile targets);

3b. *Attack of terrorist leadership* (soft targets, possibly under ground, and possibly moving; moderate air defenses); and

4. *Attack of key nodes as leading edge of military campaign.*

This set of test cases (scenarios) stresses the capabilities of candidate systems in different ways and, by and large, highlights the strengths and weaknesses of the alternatives. The quality of the testing, of course, depends on numerous details such as the quality of air defenses imputed to the adversary. The test cases should be stressful, but reasonable.

A REPRESENTATIVE SET OF OPTIONS FOR CPGS

Seven Options Analyzed

The committee analyzed seven of the very large number of possible CPGS options, informed by results of the DOD's ongoing analysis of alternatives (AoA) and the committee's independent thinking. These options were as follows:

1. *Existing systems:* A combination of tactical aircraft, cruise missiles, and long-range stealthy bombers, supported as necessary with tankers; assets for command, control, communications, computers, intelligence, surveillance, and reconnaissance (C4ISR); combat search and rescue (CSAR); and suppression of enemy air defenses. Some of these systems have very high accuracy and large payloads and, if pre-positioned in areas where targets are likely to emerge, have relatively rapid response times—but almost always measured in a few hours, not less. The exceptions are not a good basis for planning (see footnote 2 in this chapter).

2. *Conventional Trident Modification (CTM):* The option proposed in the President's budget would use existing Trident missiles on nuclear-powered ballistic missile submarines (SSBNs). The intent is to have, on each of the 12 deployed SSBNs, 2 of the submarine's 24 missiles armed with conventional rather than nuclear warheads. In its initial version, the CTM would have four kinetic energy projectile (KEP) warheads per missile, in a normal Trident reentry body augmented with an inertial navigator, Global Positioning System (GPS) receiver, and a modest maneuvering capability to strike precisely at GPS

[8]A 2005 National Research Council report entitled *Effects of Nuclear Earth-Penetrator and Other Weapons* concluded, among other things, that many strategic hard and deeply buried targets can only be attacked directly with nuclear weapons. See National Research Council, 2005, *Effects of Nuclear Earth-Penetrator and Other Weapons,* The National Academies Press, Washington, D.C.

locations on the ground. The accuracy required is quite high, and testing to confirm that it has been achieved is essential. Lethality can be achieved against soft, relatively small-area targets with diameters up to about 10 meters (more if multiple warheads are appropriately targeted across an area), and can be significant against some point structural targets, as discussed in Chapter 4. The lethality of warheads would evolve over time (e.g., with a bent-nose warhead, as described in Chapter 4, which would improve effectiveness against some terrain-protected targets).

3. *CTM-2:* An option developed in the committee's activities, this would remove the third stage of the Trident missile, thereby increasing payload capacity and conventional-payload options, while still achieving 4,000 nautical mile (nmi) ranges. Developing the CTM-2 would require relatively little modification of the existing missile, but it would require a few additional flight-tests to validate the modified guidance and control software. The base configuration would be as with the CTM—each SSBN would have two or more CTM-2 missiles. An initial version of the CTM-2 would use the CTM's KEP warheads or the large penetrator reentry vehicles (RVs) that are more effective against hard or buried targets. Even the initial version could exploit bent-nose technology, as described in Chapter 4. Later versions could include ballistically delivered unmanned aerial vehicles (UAVs) with intelligence, surveillance, and reconnaissance (ISR) packages and weapons to reacquire and attack mobile targets.

4. *Submarine-Launched Global Strike Missile (SLGSM):* Conceived by the Navy as a potential mid-term system following up on CTM experience, the SLGSM would be launched primarily from existing nuclear-powered guided missile submarines (SSGNs) but could also be launched from SSBNs to extend coverage; it would have intermediate range and a single RV payload larger than the total payload of the CTM (it could also carry CTM-like payloads, of course). Because the SSGNs have a variety of missions, there would be some operational side effects related to the use of Special Operations Forces.

5. *Boost-glide missile (initial version, CSM-1):* The Air Force has proposed the Conventional Strike Missile (CSM) based on the proposed Minotaur launch vehicle described in Chapter 4. This would have boost-glide capability, extending range and permitting large maneuvers to avoid unintended overflight. The Air Force proposed that its initial boost-glide missile have an 800-second glide in the atmosphere. The committee believes that to reduce technical risk and field a capability quickly, the initial design should be limited to the delivery of a penetrator warhead or larger KEP warheads, rather than the munitions proposed for the next option (the "second version" of CSM).

6. *Boost-glide missile (second version, CSM-2):* The Air Force proposes a second, later variant of CSM with a 3,000-second glide in the atmosphere to increase its range and maneuverability. It would have more capability, more development risk, and a longer time to deployment than CSM-1. In the Air Force concept, both the initial and second versions would have the capability to slow

enough to dispense a wide variety of the Air Force's arsenal of air-launched munitions (e.g., Small Diameter Bomb, Joint Direct Attack Munition [JDAM], and submunitions of various kinds, including the BLU-108 with Skeet warheads and possibly modules for ISR, battle damage assessment [BDA], and reattack).

7. *Hypersonic cruise missiles:* Hypersonic cruise missiles are being investigated and, if developed, could be used for CPGS if forward deployed (e.g., on SSGNs) or launched from long-range aircraft. Such hypersonic cruise missiles would have only medium range but would have considerable capability for terminal-phase dispensing of smart munitions and ISR modules, and if slowed to subsonic speeds, could execute a search to reacquire and strike moving targets.

Why Not Just Use the Minuteman?

Although the committee considered all of the many options used in the DOD AoA and invented additional options and variants on its own, an obvious question for some readers may be, Why not just use the Minuteman missile? The committee investigated the option, but concluded that, although technically viable, the Minuteman III is not a realistic contender for reasons involving the complexity of issues described in Appendix I of this report. In short, the required renegotiation of the Strategic Arms Reduction Treaty, combined with "not-in-my-backyard" issues, presents substantial challenges to deploying intercontinental ballistic missiles (ICBMs) in locations such as Hawaii or Guam, particularly in the near term, as compared with other challenges associated with CTM deployment. Furthermore, reducing the number of conventionally armed ICBMs to be deployed in Hawaii or Guam or using mobile units rather than silos will not offset these concerns. Like the Air Force, the committee gauged the pros and cons of land-based conventionally armed ballistic missile systems and decided against recommending the Minuteman as a near-term option.

ANALYSIS

Functional and Technical Analysis

With the foregoing as background, the committee's independent analysis of CPGS alternatives follows. Time and resource constraints precluded an in-depth study, but the analysis provides useful insights.

Table 2-2 summarizes a qualitative technical comparison of the options, focusing only on military-technical issues. (The knotty "nuclear ambiguity" issue identified by some observers, including some in Congress, is discussed in Chapter 3.) Table 2-3 summarizes the committee's rough estimates of *earliest* availability and total 20-year system costs. The estimates are informed by material from the DOD and from contractors and by some committee members' long experience in capability development.

TABLE 2-2 A Qualitative Technical Comparison of the Conventional Prompt Global Strike (CPGS) Options Reviewed by the Committee

	Investment Options							
Measures	Existing Systems	CTM	CTM-2	CTM-2 Variant (with UAV)[a]	SLGSM	Boost-Glide Missile (CSM-1)	Boost-Glide Missile (CSM-2)	Hypersonic Cruise Missiles
Execution time								▲
Positioning/final planning time								
Geographic coverage								
Defense penetration							▲	▲
Lethality		▲						
Ability to attack moving targets								
Volume of fire								

Assessment and reattack	Marginal/Mixed	Very Poor	Very Poor	Very Good	Very Poor	Very Poor ▲	Marginal/Mixed ▲	Marginal/Mixed ▲
Controllability	Good	Very Good	Very Good	Very Good	Good ▲	Very Good	Very Good	Very Good
Impact on other missions	Very Good	Good	Good	Good	Poor	Very Good	Very Good	Very Good
Need for bases, overflight	Very Poor	Very Good	Very Good	Very Good	Very Good	Very Good	Very Good	Very Good
Earliest availability (initial operational capability)	Very Good	Very Good	Good	Very Poor	Marginal/Mixed	Poor	Very Poor	Very Poor
Development risk	Very Good	Very Good	Good	Poor	Good	Marginal/Mixed	Very Poor	Very Poor
Confidence in high reliability	Very Good	Very Good	Very Good	Poor	Good	Good	Poor	Poor ▲
Value for capability evolution	Poor	Very Good	Very Good	Very Good	Marginal/Mixed	Very Good	Very Good	Very Good

▲ A warning marker (top right corner of cell) indicates a result of averaging good and bad cases.

Very Good | Good | Marginal/Mixed | Poor | Very Poor

NOTE: A black warning marker indicates a result of averaging good and bad cases, for example, cases with and without advanced air defenses. Scores for many rows assume fully successful development and political circumstances permitting CPGS; they do not consider countermeasures and counter countermeasures except where mentioned in text.

[a]CTM-2 with unmanned aerial vehicle (UAV) is not among the seven CPGS alternatives reviewed by the committee; however, it is included in this table as a postulated variant of CTM-2 to indicate its potential value against more difficult mobile targets.

TABLE 2-3 Estimated Earliest Initial Operational Capabilities (IOCs) and Costs

Alternatives	Earliest IOC	20-Year Cost (relative to CTM: billions of 2009 dollars)	Cost to IOC (relative to CTM: billions of 2009 dollars)
Existing systems	Available now	Not Applicable	Not Applicable
CTM	2011	1	0.5
CTM-2	2013	3	1.5
CTM-2 (with UAV)	2016-2020	5	3.5
SLGSM	2014-2015	5-10	5
Boost-glide missile (CSM-1)	2016-2020	10-20	5-15
Boost-glide missile (CSM-2)	2018-2024	10-25	5-20
Hypersonic cruise missiles	2020-2024	10-20	5-15

NOTE: Actual IOCs for all but the CTM are likely to be later for many reasons, including delays in decision making, the time required to stand up program offices, and other factors. The costs shown are merely rough estimates to make distinctions. Actual costs are more likely to be higher than lower, but the committee believes that the ordering is about right. Acronyms are defined in Appendix A.

In Table 2-2, the color coding indicates the committee's assessment of options as "very good," "good," "marginal/mixed," "poor," and "very poor" (or equivalents, such as a very high development risk or late initial operational capability [IOC] being treated as "very poor"). It is essential that the reader recognize the ground rules for the assessment to avoid misunderstanding. The most important are discussed below.

- In most rows of Table 2-2, the committee evaluates the options with the assumption of success; that is, the option is scored for the measure in question (e.g., execution time) based on the planned capability. However, in other rows (those near to the bottom), options are marked down because of high development risk, long development time, or questions about whether the systems would have high reliability.
- The committee also evaluates the options on a relative basis, looking at what is being sought for the measure in question. For example, an execution time of 1 hour would rate as "very good" even though the committee might prefer zero time of flight. More subtle, a "very good" lethality score does not, for example, imply capability against large-area, hard and deeply buried target complexes about which there is no detailed information on vulnerabilities. Taking an example at the other end of the spectrum, a very poor score (such as for execution time) does not mean that the option would *never* be able to execute quickly, but rather that it could do so only in a very restrictive set of circumstances.

- Some evaluations are inherently ambiguous because they are an "average" across cases with markedly different results. Cells evaluated as "marginal/mixed" with warning markers in their top right corners indicate evaluations that could be either good or bad, depending on unspecified details, or cases on which there were mixed evaluations.

It follows that assessments such those in Table 2-2, and in Table 2-4 on pages 48-49, should not be interpreted out of their analytical context. Most important, they should not be read to mean that any of the options would never have capability or that any option would always have capability.

With these prefacing comments, the basis for the committee's assessments can be described briefly as follows.

1. *Execution time.* For reasons discussed earlier, a goal of 1 hour for execution time was used as a reasonable basis for scoring. Existing systems (tactical aircraft, bombers, and cruise missiles) rate poorly because they could achieve such times only in very special cases (see footnote 2 in this chapter). Moreover, if the targets at issue were deep inside a large country, far from launch locations, execution times might be much longer (e.g., 10 to 20 hours, depending on launch point). Hypersonic cruise missiles would depend on forward-stationed forces, and boost-glide missiles such as the CSM-1 and CSM-2 are intended to have longer flight distances and flight times, which might well be adequate but could be short of the 1-hour goal.

2. *Positioning/final planning time.* Even if forward stationed and poised, existing systems may not prove to have been stationed and poised in the right place for a strike and may require support by tanker aircraft and other enablers. Further, final mission planning can take considerable time, especially for working out paths of flight to avoid or defeat air defenses.[9] This would more likely afflict existing systems or hypersonic cruise missiles than it would affect dedicated CPGS systems. These options, therefore, are marked down slightly because they depend on special circumstances.

3. *Geographic coverage.* Most of the options will have the geographic coverage that the committee believes necessary. The exception for existing forces, SLGSM, and hypersonic cruise missiles is targets deep inside large countries. Hence they are scored down somewhat on this measure. This is an instance in which the committee has not required full global reach.

---

[9]Another factor in the time line, of course, has to do with "enabling" actions, discussed later in the chapter. These involve gathering and assessing intelligence, establishing precise geolocations of targets, decision making, and transmission of orders. Every effort should be made to do whatever is possible to reduce these other elements of the time line, even though they are often not the limiting factors.

4. *Defense penetration.* Advanced air defenses are a threat to existing systems and are likely to be purchased by states that are far less capable generally than a "peer competitor." There will continue to be cases—in particular as stealth technology and countermeasures evolve—where defenses can be defeated, but in other cases they would severely affect the feasibility *or risk* of missions (e.g., the risk of having crew killed or captured). U.S. ballistic missiles are likely to have minimal penetration problems for many years (although countermeasures might be necessary). Thus, the CTM, CTM-2, SLGSM, and CSM-1 (initial version of the boost-glide missile) all merit "very good" scores. Evaluations of the other systems are difficult to make, and differences of opinion exist even among experts. Some studies suggest that, because of their unpredictable maneuvering, these systems should be able to penetrate advanced air defenses. Some experts on the committee were less sanguine because, if used with some payloads, CSM-2 (the second version of the boost-glide missile) and hypersonic cruise missiles must slow down substantially, either to dispense munitions, to conduct ISR activities, or to assess and reattack. That would increase their vulnerability to air defenses. This said, such payloads would arguably be most relevant to the attack of targets that would not be well defended (e.g., terrorist leaders). Because of such ambiguities, disagreement, and case dependence, the committee rated the CSM-2 and hypersonic cruise missiles "marginal/mixed," with the warning markers in Table 2-2 indicating ambiguity (i.e., the systems can be good or bad by this measure, depending on details of circumstance).

5. *Lethality.* The baseline version of the CTM, assuming that it achieves its accuracy specifications, is expected to be highly lethal for some important targets, but there is a significant set of potential targets—mobile, hard, or extensive—against which it will have limited capability, not only compared with more ambitious CPGS concepts but compared with existing systems. This is due to the CTM's small reentry bodies and—in the baseline concept—small number of weapons available per target, as well as to the nature of the warhead. The CTM's lethality can be improved incrementally in ways that the DOD recognizes and in additional ways that the committee suggests (see the discussion later in this chapter), but the CTM is not credited with those improvements here. The other systems should have greater lethality against a range of targets, assuming necessary intelligence—and, in some cases, an ability to receive and use target location updates after launch. Some of the weapons systems would have a variety of loading options (e.g., penetrators, area weapons). No CPGS option is likely to be effective against all targets, even with good intelligence. For example, hard, deeply buried target complexes are very difficult to attack without knowing their special vulnerabilities. Thus, a "very good" score does *not* mean high lethality in all cases; it means only that a reasonable goal for CPGS weapons has been met.

6. *Ability to attack moving targets.* Capability against moving targets would be minimal for all options unless they had the ability to detect and track such targets or to receive and use updated information from other platforms until very

late in their flight paths. Such capabilities are postulated for CTM-2 (with UAV), for CSM-2 (the second version of the boost-glide missile), and for the hypersonic cruise missiles, although achieving those capabilities entails high development risk and long development times. Further, the sufficiency of capability would be in doubt given the potential for adversary tactical countermeasures (e.g., proliferation of targets) and dependence on intelligence to narrow search areas for submunitions.

7. *Volume of fire.* The number of weapons that might be employed in a CPGS mission is perhaps in the range of 10 or fewer for scenarios such as those involving terrorist leaders or WMD shipments and perhaps as many as 100 (which would require a CPGS deployment larger than those considered in this study) in scenarios employing CPGS as the leading edge of a larger military campaign. The CTM would have a limited volume of fire (although it could be increased by using more warheads per missile, at the expense of some range), so it is marked down somewhat.[10] The other systems have sufficient volume of fire for the missions on which the committee assessed capability, which might require tens, but not hundreds, of weapons.[11]

8. *Assessment and reattack.* For some of the options, it is possible for the strike system itself to conduct a degree of assessment and/or a degree of BDA, which could be crucial in subsequent military decisions, including those about the need to reattack. Some of the options, particularly the CTM-2 (with UAV), CSM-2 (the second version of the boost-glide missile), and hypersonic cruise missile options, envision releasing sensors and communication devices as part of the strike; these have the potential to support real-time BDA and reattack of some targets as part of the strike itself, although achieving such capabilities is a formidable longer-term challenge.

9. *Controllability.* It is important that CPGS systems be subject to very high levels of safety and security. It appears that the currently planned control mechanisms vary with the option. They would be quite stringent for the CTM

---

[10]The committee considered only limited deployments of CPGS systems. Large-scale deployments would raise very important issues, such as the appearance of a disarming first-strike capability, that have not been addressed. The committee recognized that the CTM proposal involves only two conventionally armed missiles per submarine, and did not analyze or discuss in detail the issues that would arise from the deployment or use of larger deployments.

[11]It is sometimes suggested that CPGS systems could have a role along with nuclear weapons in planned options for strategic attacks, by using very long range, highly accurate conventional strikes in place of nuclear weapons to hit suitably soft and/or small targets that are now assigned to nuclear weapons. This concept of CPGS raises different issues from those posed by either the "niche" capabilities or the "leading-edge" capabilities discussed here. Among many other differences, the numbers required would almost certainly have to be quite large, if the stated purpose of achieving Single Integrated Operational Plan (SIOP)-like effects while using markedly fewer nuclear weapons were to be achieved. (The SIOP is a classified blueprint that specifies how nuclear weapons of the United States would be used in the event of nuclear attacks.)

and CTM-2,[12] but perhaps less so for the others, especially the SLGSM launched from SSGNs. One issue for the SLGSM is whether a weapon could physically be launched without the conventional-weapon equivalent of an emergency action message. If the answer is no, that is, if the same protections would be built in as for the CTM, then the SLGSM's score by this measure would be "very good."

10. *Impact on other missions.* A significant question raised by the Congress is whether the adoption of a given CPGS option would compromise other missions. The CTM program and the committee's baseline configuration for the CTM-2 involve a mixed load of nuclear and conventional missiles. It has been argued that this would create operational problems and also lead to new operating procedures that might degrade somewhat the survivability and target coverage of the SSBN force. The DOD has studied such matters, and the membership of the committee included individuals with personal command experience in such matters. The committee concluded that these adverse consequences would be modest and quite manageable. They would be larger if the CTM or CTM-2 were deployed in a "pure-load" configuration (i.e., perhaps two Trident submarines, each carrying 24 CTM or CTM-2 missiles and no nuclear weapons). Such an approach might reduce the ambiguity concern, but it would have a larger impact on nuclear operations because of the reduced number of nuclear-armed SSBNs (unless the size of the fleet were increased, which would be expensive). It is the committee's judgment that the number of available nuclear weapons could be maintained by increasing the number of warheads on those missiles with nuclear weapons, but there would still be concern about having concentrated the on-alert nuclear submarine-launched ballistic missiles (SLBMs) on two fewer on-patrol submarines.

The biggest operational impact of CPGS would be associated with the SSGN-based SLGSM option. The primary impact would be due to the SSGN's having important SOF-related missions,[13] which would tend to call for different operational deployments and operating procedures than would the CPGS mission. The SSGN crews would have to train for the different missions, and choices would need to be made, perhaps frequently, about relative priorities.[14]

11. *Need for bases, overflight.* One of the primary features of the CPGS concept is the capability to permit an immediate (less than 1 hour) strike. Although the United States has forward bases and arrangements with local states permitting or tolerating overflight rights, and although unilateral overflights can be made

---

[12]Ironically, this level of security and safety, including the use of emergency action messages, is regarded by some as increasing the so-called nuclear ambiguity problems addressed later in the report.

[13]For example, SSGNs may be charged with the transport of Special Operations Forces, a mission that may be compromised if the focus is instead on CPGS readiness.

[14]Precisely which SSGN missions would be emphasized at a given time is unclear. A Congressional Budget Office study discussed the range of potential missions that have been mentioned, showing the potential for conflicts (Congressional Budget Office, 2002, *Increasing the Mission Capability of the Attack Submarine Force,* Washington, D.C., March, pp. 16ff.).

when necessary in any case, these considerations are weighty and may reduce the U.S. ability to act quickly and unilaterally.[15]

12. *Earliest availability (IOC).* It is notoriously difficult to estimate the time at which a weapons system would have initial operational capability (IOC), or final operational capability (FOC). That difficulty increases with the ambitiousness of the program at issue. Program offices and contractors have incentives to be optimistic, engineers think in terms of what *could* be done. Others worry about what is *likely* to be done, given the probability of delays associated with decision making in both the DOD and the Congress, funding slippages, the time required to set up or reinvigorate development organizations and commands, and the record in recent decades of IOCs coming later than initially advertised.

The committee saw contractor briefings suggesting that almost all of the CPGS options could be fielded within about 5 years from go-ahead (and at remarkably low costs). However, after reviewing the materials provided by the program offices of the Navy and Air Force, the Office of the Secretary of Defense (OSD) (Acquisition, Technology and Logistics), and the U.S. Strategic Command (STRATCOM), as well as the analysis of alternatives being conducted for the DOD with the Air Force as executive agent, the committee—drawing also on the judgment of its own members with relevant experience, expertise, and skepticism—estimated the IOCs as shown in Table 2-3. These are *earliest* plausible IOCs. The CTM's IOC is highly credible (subject primarily to early authorization) because it is a very incremental effort within a mature organization with a long track record of delivering; the CTM-2's earliest IOC is shown as a year or two later because it seemed that *any* change of concept would involve some delay, and because some testing of the two-stage Trident with a CTM payload would be needed. The IOC shown for the SLGSM is also credible, since the Navy has laid much of the technical groundwork, would again be pursuing incremental development within an existing organization and has been consistently conservative in its estimates. Initially, the committee believed that the boost-glide missiles or hypersonic cruise-missile options were likely to be long term in nature, more like 2020 than earlier. It is *possible* that research and development (R&D) tests that are already programmed will prove successful and reduce some of the more worrisome risks—for example, that of the ability of materials to tolerate long hypersonic periods of flight. It is also possible that decisions could then be made with unusual swiftness to authorize full programs. Thus, it would then seem *possible* for IOCs to be achieved as early as 2015, although not with the same level of performance reliability and stability as would be possible for the CTM, CTM-2, and SLGSM options. Accordingly, the committee's estimate of the earliest IOC of each of these boost-glide missile and hypersonic cruise-missile systems is

[15]This criterion should not be misinterpreted. Forward basing and other access arrangements are desirable and important for many strategic reasons that would not be affected by the existence of CPGS capability.

expressed as a range. In summary, the committee concludes that there are only two relatively early options (both based on Trident missiles and submarines), one highly credible mid-term option, and two (more, if one counts variants) interesting options with mid-to-long-term potential.

13. *Development risk.* The CTM and CTM-2 options clearly have the lowest risk and, based on empirical knowledge and the modest level of technology extrapolation, the committee judges the risk to be low on a relative basis—assuming, however, completion of the careful testing program that is planned. The successive options involve increasingly more risk. The committee's assessments reflect systematic consideration of how much of an extension of current technology is involved relative to prior capabilities, technology readiness levels (TRLs) for component technologies, and other factors such as maturity and past success of industrial and government organizations.

14. *Confidence in high reliability.* In principle, any of the systems under consideration could be designed to achieve high reliability. However, the committee noted that strategic systems have been designed for high reliability (more than 90 percent reliability) routinely for decades, whereas tactical systems have widely varied reliabilities—in part because it is difficult and expensive to raise the reliabilities of air-breathing systems to high levels. For this reason, the committee expressed doubts about the confidence that would be achieved in the reliability of CSM-2 or the hypersonic cruise-missile options, at least at the time of IOC. Further, the committee reasoned that the UAV-carrying variant of the CTM-2 would be more likely to have early difficulties, even if largely successful. And, finally, there are enough differences between the SLGSM and CSM-1 on the one hand and existing systems on the other that difficulties might be expected. This kind of qualitative reasoning was the basis for the evaluations shown.

Even if developments proceed well, the committee believed that the reliabilities achieved in the timescales noted might prove unacceptably low for the CSM-2 and hypersonic cruise-missile options. Better results would then be achievable in time. It *may* be that reliabilities would be high from the outset, at least for the boost-glide missile (initial version), but risks for the CSM-2 and hypersonic cruise-missile options would have to be judged to be considerable. Again it is noted, however, that the issues here are not fundamental; higher reliabilities could be required and achieved with sufficient effort.

15. *Value for capability evolution.* A measure that the committee believed to be important was whether a given option had growth potential or would contribute materially to the development of follow-on systems. Particular follow-on systems might or might not be desirable and, as discussed in Chapter 3, there are profound issues involved with such development and deployment decisions. Nonetheless, from the pragmatic perspective of investing in system development, evolutionary *potential* is important. The assessments presented bear explanation. The CTM itself would provide not only an important early capability but also an important basis for further development, either to the CTM-2 or to something

like the SLGSM. It would provide considerable technical and operational data and experience relevant to the boost-glide option. The CTM-2 would have even more evolutionary potential because it would have a larger payload capacity. The SLGSM, as described, has more limited evolution potential because the missile has only intermediate range and limited capacity for growth to larger payloads. The boost-glide missile and hypersonic cruise-missile options would have plenty of room for further evolution, although concerns about air defense are significant, and some committee members are skeptical about the feasibility—even in the long run—of being able to use such systems effectively against noncooperative moving targets.

Effectiveness Analysis

The committee conducted a rough analysis of the effectiveness of the planned capabilities of various CPGS options using the spanning set of scenarios described earlier. This analysis, which is based on the four major operational scenarios covered by the test cases, is summarized in Table 2-4. Cells in this table with a warning marker in the top right corner have been evaluated as a compromise between better and worse cases. That is, depending on details of the scenario, the system might be very effective or ineffective. The factors include the presence of air defenses, whether targets are moving, and whether the payload involves penetrating weapons or dispensed weapons.

Discussion of Evaluation in Test Cases

The test cases are formulated specifically for the purpose of evaluating the potential of systems for CPGS in the period between 5 and 15 years from now, not for global strike generally, and not for today.

Existing systems do not do well in the committee's evaluations because of not being "prompt" enough and because of worries about emerging air defenses and/or inability to attack moving targets. This is so even though there are circumstances where existing systems could be used effectively, as they have in the past. The problem with existing forward-deployed strike forces is that they are generally not close enough to execute a strike within 1 or 2 hours—even within a theater of operations. It takes time and diverts many resources to generate forces and supporting functions for a strike. Tanking must be provided and coordinated with strike tactical aircraft, and often overflight permission or defense penetration must be dealt with. Mission planning can be time-consuming when air defenses are a problem. For these reasons, these forces get low marks for execution time as it affects the scenarios in this assessment. Finally, while the development and operating costs and risks are minimal, given that they already exist and are deployed forward anyway, relocating such forces entails a large expense because of the large logistics tail that supports them.

TABLE 2-4 Evaluation of the Conventional Prompt Global Strike (CPGS) Options in a Set of Test Cases

	Measures			
Investment Options	Respond Fast to Attack or Threat	Attack Weak Points of Hardened WMD Facility[a]	Attack Terrorists Fast (Leaders or WMD)	Strike C3 Nodes in Leading-Edge Attack
Existing systems				
CTM				
CTM-2				
CTM-2 (with UAV)				

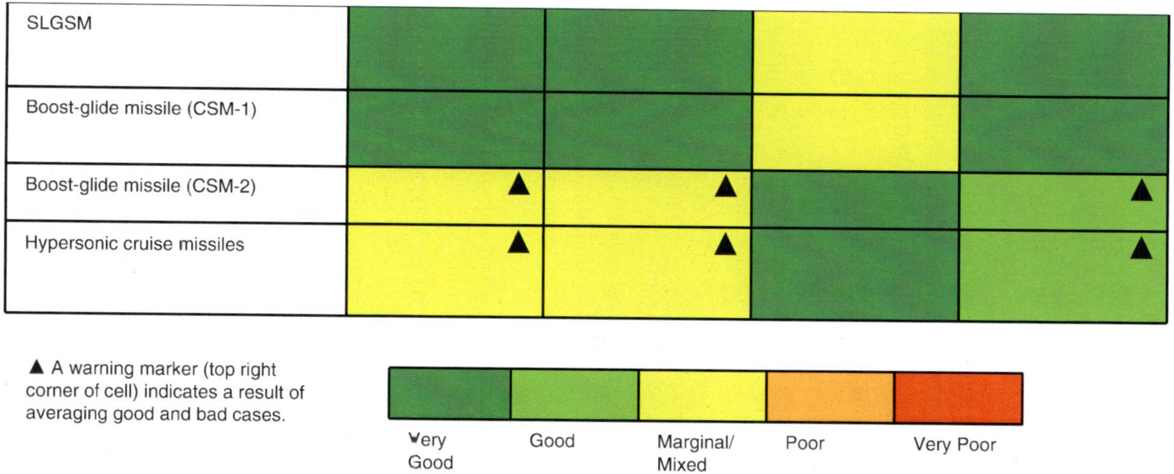

NOTE: Evaluations assume successful development and political circumstances allowing use of prompt global strike. They do not consider countermeasures or counter countermeasures. Evaluations also assume "good cases." For example, none of the weapons would have significant capability against large, complex, hard and buried complexes—unless the complexes had special vulnerabilities that could be exploited. Similarly, moving targets would only sometimes be targetable.

Acronyms are defined in Appendix A.

[a]An effective CPGS capability against hard and deeply buried targets is restricted to special cases. In this column, the selected cases assume advanced intelligence about the targets and their vulnerable accoutrements, such as their entrances, and the inability of the facilities to function effectively if the accoutrements are disabled or destroyed. Furthermore, it is important to note that a 2005 report of the National Research Council entitled *Effects of Nuclear Earth-Penetrator and Other Weapons* (The National Academies Press, Washington, D.C.), concluded, among other things, that "[m]any of the more strategic hard and deeply buried targets are beyond the reach of conventional explosive penetrating weapons and can be held at risk of destruction only with nuclear weapons" (p. 1).

The CTM succeeds in the test cases shown in Table 2-4 (columns one and four) because of its promptness, but it has no ability to attack hard facilities, only limited capability against terrain-protected targets, and no ability against moving targets.[16] The score for the "Attack Terrorists Fast (Leaders or WMD)" test case (third column in Table 2-4) is mixed because the targets might or might not be moving. In the latter case, the CTM and CTM-2 would be capable. Given a UAV, the CTM-2 could have capability against moving targets as well. The SLGSM and CSM-1 have no moving-target capability and so are rated the same as CTM and CTM-2.

The CSM-2 (with UAV) and hypersonic cruise missiles would have moving-target capability, but, depending on load and possible maneuvers, might or might not be able to cope with advanced air defenses. Such defenses would probably not apply for the "Attack Terrorists Fast (Leaders or WMD)" test case, however. For the other cases, the table shows results as "good," but "marginal/mixed." These systems should do relatively well if loaded with payloads not requiring them to slow up in the terminal phase. With payloads requiring the slow-up, it is unclear whether they would be able to deal well with advanced air defenses. It is a matter of speculative disagreement. On balance, this "Attack Terrorists Fast (Leaders or WMD)" test case seemed to merit a score of "good."

By and large, all of the planned options would achieve significant capability in most of the test case scenarios. The moving-target cases pose the most serious challenge. It is imperative to understand, however, that the evaluations are for "good cases," such as when a deeply buried target has a special vulnerability, or when moving targets are out in the open where collateral damage is not a concern. None of the options would have capability in all versions of any of the scenarios, but the analysis indicates that the "good cases" are plausible enough to plan for.

One of the factors complicating this analysis is that effectiveness depends on weapons loading, and most of the options have multiple potential loadings. The DOD will want more explicit analysis regarding the ability to adjust loads (and concepts of operations) based on strategic warning. For example, if a plausible threat to nuclear SSBNs arose, SSBNs could return to maximally secure operations; or more likely, if the need to attack hardened sites became evident, unitary penetrator weapons might be favored over KEPs. How long would it take for reloading? Some weapon systems could have a mix of weapon types at a given time; others could not.

[16]The committee evaluated the CTM as it was described to the committee, but noted that the CTM could be given more capability against terrain-protected targets relatively easily and that eight, rather than four, RVs could be carried on a given missile, thereby increasing the volume of fire and capability against area targets or targets for which geospatial location information was slightly unreliable. Flechette density could also be increased, with more numerous, smaller, flechettes.

ENABLERS OF CONVENTIONAL PROMPT GLOBAL STRIKE

So far this chapter has focused on the potential effectiveness of the CPGS platforms and weapons. It has pointed out that in many and perhaps most of the plausible circumstances of CPGS employment, there would be warning time permitting intelligence collection and synthesis, precise targeting, preliminary decision making, and other preparations. In such cases, the measures discussed (e.g., execution time, lethality) would be especially important. In some cases, however, many of the "enabling activities" would also need to be done quickly—if not in minutes, then in hours or a few days, rather than weeks and months. Whether or not additional time was available for them, *the enabling activities would be critical, not merely desirable*. The material that follows touches on such enablers, including the challenge of ensuring that aimpoints are properly geolocated.

Challenges for Enablers

The use of long-range missiles to deliver conventional weapons accurately enough to damage targets requires information that is significantly more difficult to provide than that needed for nuclear weapons or for conventional weapons delivered by aircraft or short-range missiles. The primary issues relate to the following: (1) command and control (C2) that reserves decisions to the National Command Authority, while delivering information to the weapon system quickly for a short overall execution time; (2) the provision of the information necessary for accurate weapon delivery to a specified aimpoint; (3) the accurate location of aimpoints; and (4) target detection.

The following paragraphs discuss these issues in order. At the outset it is noted that the timely information available for early versions of CPGS, such as CTM, will be sufficient for many important targets of interest. However, it will be some years before the information will be sufficiently timely for a broader set of targets, especially those that are moving while being attacked. That is, one should assume considerable evolution over time and consider the possibility of following a dual track; deploy quickly a system able to expand the present capability and use existing technology (including technology for enablers) while pursuing work toward longer-term capabilities as technology advances (and, possibly, new missions are identified).

Command and Control

The command-and-control problem for the CTM is greatly simplified by adapting modestly the existing C2 system used for the nuclear-armed Tridents—the system developed for controlling the major strategic nuclear system of the nation. Its CTM variant[17] would be both effective and cost-effective. Such is not

[17]The proposed program would add a number of hardware and software provisions to prevent

necessarily the case for alternative future systems, which are being conceived as strictly conventional, in part to get away from the concerns expressed by some observers about mixed loads of weapons (i.e., mixes of nuclear and conventional). Such future systems, as currently conceived, would not be tied into the nation's C2 system for strategic (albeit non-nuclear) weapons, even though decisions about the use of such strategic weapons will almost surely be made at the highest level of the government. It is unclear to the committee whether physical control of the weapons will be strong enough, and whether the mechanisms will exist for an efficient matching of targets and weapons—something routinely considered in the existing nuclear C2 system. The committee believes that any CPGS option should be regarded as "strategic" and demanding of extremely high safeguards.

Accuracy

The accuracy required for CPGS is on the order of meters in each of three dimensions (two dimensions on the surface of the ground and one "vertical" dimension relating to target elevation and fusing altitude). Such results have been achieved with short-range systems after several iterations of guidance-system details (in Chapter 4, see the subsection entitled "Guidance, Navigation, and Control Accuracy Issues"), but the CTM or other future systems will need to achieve excellent accuracy from the outset. Even with results from testing, ensuring that a single or a handful of shots will achieve such accuracy every time is very challenging. Several studies have shown that bias errors will be troublesome in some cases for the CPGS systems.[18] Given the limited payloads and small radius of effects for conventional weapons, there is little room for error in any of the three dimensions. These CPGS systems will require GPS updates to correct inertial guidance systems toward the end of flight and perhaps will need to include additional systems to correct for bias errors (a capability not available for early systems). Also, the requirement to protect the weapon itself from heat and other atmospheric effects implies correct compensation for aerodynamic effects over significant time periods, during which velocities and orientations are changing significantly. Development and full-scale, system-level testing are essential. That testing is one of the highly attractive features of the proposed CTM program and would be important whether or not the CTM were deployed. If CTM research,

launch of a nuclear weapon rather than a conventional weapon when a conventional-weapon attack was intended. Safety, in this case, would rely not only on rigorous procedures but also on well-tested hardware and software safeguards. The committee found this aspect of the CTM's design to be especially attractive.

[18]Accuracies are usually quoted in circular error probable (CEP) terms, with CEP measured around the intended impact point. Bias errors are another matter: the intended impact point may be displaced significantly from where the actual target is located. One contributor to that error is target location error (TLE), and it is not a function of the guidance system or the number of times that the missile has been flight-tested.

development, testing, and evaluation (RDT&E) are terminated, other systems will start from a much more primitive technological base.

Aimpoint Location

Also challenging is determining the location of the target to be struck. CTM and its early evolutions will be limited to striking locations known before launch—locations that contain the right targets and have reliably established coordinates. Targets will need to be stationary for at least 1 hour to be struck effectively, unless their arrivals at fixed locations can be accurately predicted. Future versions of the weapons might allow an in-flight update of the target location, but the targets would still need to be stationary for the times between last update and impact, which might be tens of minutes.

Any error in the actual location of the target (target location error, or TLE), including bias errors, needs to be significantly smaller than the weapon's range of effects, which means that TLE needs again to be on the order of a few meters. While such accuracy is difficult in the two dimensions of ground position, it is even more difficult currently for the vertical dimension. That situation is improving, as discussed below.

Horizontal accuracy can be improved significantly by measuring position relative to a known "fiduciary point" recorded by the National Geospatial-Intelligence Agency (NGA). NGA produces an extensive Digital Point Positional Data Base (DPPDB) of reference photos with embedded and accurately determined reference points, or fiduciary points. The DPPDB enables mission planners to derive in minutes accurate geographic coordinates for a visually observable point—for example, from a recent photo taken in the field. The DPPDB today covers significant areas in known hot spots around the world. In the event that a target emerges in an area of the world not covered by the DPPDB, the process of reducing TLE to an acceptable level can take days or even weeks. In cases where the TLE is only somewhat degraded, CTM with KEP warheads could still be effective by pattern-targeting its RVs over the target to increase the area over which lethal effects are spread.[19]

Target Detection and Localization

In addition to the problems of location and control described above, there remains the challenge of finding a target in the first place, especially in the case of a movable, but not continuously moving, target at rest. If U.S. forces are deployed in the area and maintain control of the airspace, a high-altitude system such as

[19] "Pattern-targeting" as used here means that the individual shots would be directed at slightly different aimpoints in a pattern around the nominal aimpoint so as to increase the probability of a successful impact despite small residual bias errors in the target's actual location.

Global Hawk could detect and locate a target with synthetic aperture radar or light imaging—assuming suitable fiduciary points. However, where satellites are needed to detect the target, the detection will be episodic at best. And sometimes it will be necessary to rely on human intelligence and to translate information like "The ship is moored at Pier 9" into coordinates. Today, with the war in Iraq requiring maximum support by scarce ISR resources, the competition for satellite resources is severe, but it is difficult even in more normal times. There are significant limits on how quickly satellite tasking can be adjusted to accommodate even the highest intelligence priorities. Fortunately, when there is activity somewhere in the world that might call for use of a CPGS weapon, there is likely to be heightened interest for other reasons as well, so that the satellite systems will be concentrating their coverage on the problem area.

The National Reconnaissance Office (NRO) developed plans and programs in the 1990s to address the revisit problem by increasing the area of Earth observed per day by a factor of about 100. While there have been program cancellations and delays, about a factor-of-10 improvement should have been effected by the time the CTM is deployed. The planned Air Force Space Based Radar Program, which is in the initial stages of implementation, will provide relatively persistent coverage of chosen points on Earth, with maximum gaps between available observation times measured in tens of minutes.[20] This revisit time does not guarantee the ability to track a target with certainty to see where it stops, but it is effective where congestion is not too high, such as in areas where potential CTM targets would most likely exist. If this program is deployed, by 2020 its ability to find targets for CTM and its successors will change from being episodic to being relatively reliable. In addition, these new systems (the NRO programs and Space Based Radar Program) will allow acceptable TLE determination after a few observations—after approximately 30 minutes with systems of the Space Based Radar Program and within a day or less for those of the new NRO programs, without using fiduciary points, if this capability is built into these systems.

Only when sufficient work has been accomplished in advance to provide very accurate fiduciary points near a target will CTM and its successors be provided with information about fixed aimpoints with a TLE small enough to expect that the weapon will achieve the programmed damage. A patterned attack with multiple RVs could mitigate these problems. Adding more RVs to the CTM, as discussed elsewhere in this chapter, would increase the ability to do such patterned attacks. For fixed facilities, this is straightforward. For other potential target locations that have a time-dependent value, such as a pier that is a valuable target when it has WMD cargo ready to load but not when it is empty, the detection of the

---

[20]Congressional funding for development of the Space Based Radar Program has been limited over the past few years. For an overview of the program, see General Accounting Office, 2004, *Defense Acquisitions: Space-Based Radar Effort Needs Additional Knowledge Before Starting Development,* GAO-04-759, Washington, D.C.

increased value will probably be provided by satellites or other sources episodically. Because CTM and other envisioned CPGS systems will rely on ballistic or high-altitude delivery of the weapon, terrain masking and other protection capabilities will be relatively ineffective, and additional data concerning the surrounding area will most likely not be needed. However, when payloads must be delivered using maneuvering RVs with limited corner-turning capability, the requirement for information to support flight planning will increase. Such data should still be available from the NRO systems and would be available from the space-based radar when the payloads are delivered by canister into what is in effect an atmospheric-flight terminal engagement. The requirement for accurate position location is then reduced significantly, as is the constraint on effectiveness on mobile targets, because that engagement can use onboard sensors to control terminal maneuvers—for example, GPS plus fine imagery to move the aimpoint by a few meters.

Because the initial CTM system will have so few system tests before deployment, the tolerance for error in the enabling systems will be small. It will be very important to have tests by the NGA using precise measurement of errors in the location of known fiduciary points to detect and overcome bias errors. In addition, NGA will need to determine accurately and quickly fiduciary points in the vicinity of potential targets to use when a satellite happens to locate a target of interest.

Later versions of CPGS could use a more robust system of enablers, with world-networked capabilities to share information being deployed, and significantly improved satellite reconnaissance capability to find targets and track them. Later CPGS versions themselves will have better terminal systems and will also require less external information in order to be effective. Exactly how those capabilities will be combined is unknown at this time, but the constraints on the CTM itself are well defined and subject to analysis, simulation, and, most importantly, testing of the enabling systems.

Ensuring Safety and the Absence of Errors

A strategic system such as CPGS must be designed with extremely high levels of protection to ensure safety and the absence of mistakes. As the world was reminded recently when nuclear-armed missiles were accidentally loaded onto a B-52, transported to another air base, and essentially left unattended for many hours, mistakes can happen—even with the nuclear weapons to which the United States has applied special precautions for decades.[21]

[21]In response to the incident of unauthorized movement of nuclear warheads from Minot Air Force Base, North Dakota, to Barksdale Air Force Base, Louisiana, Secretary of Defense Robert Gates authorized an independent investigation of the incident resulting in the following report: Defense Science Board, 2008, *Report on the Unauthorized Movement of Nuclear Weapons,* Permanent Task Force on Nuclear Weapons Surety, Washington, D.C. For additional reading, see Joby Warrick and Walter Pincus, 2007, "Missteps in the Bunker," *Washington Post,* September 23.

The proposed CTM program raises at least the possibility of an accidental launch of a nuclear weapon instead of the intended launch of a conventional weapon because (1) both kinds of weapons would be carried on the same submarine platform and (2) prompt global strikes may often allow little time for second checks.

General Considerations

The committee discussed at some length this matter of the need to ensure safety and the absence of errors with respect to CPGS, drawing in part on the knowledge of several members who have been involved for many years with strategic command and control. Some observations follow.

1. *Ensuring safety and security is a system problem.* Errors can occur in any of many parts of the overall system for handling and employing weapons: for example, during production, handover to the military customer, transport to a military base, warehousing, selection of ordnance to be loaded on an operational platform, handling within the platform, the order to use a particular type of weapon for a particular target, the choice of a weapon consistent with the command order, and the delivery of the weapon to the correct target. And, when weapons are returned to storage, decommissioned, and destroyed, additional opportunities arise for problems. The events that led to the recent incident involving the unintended transport of nuclear weapons on a B-52, followed by many hours during which the missiles sat unattended on the airplane at Minot Air Force Base, North Dakota, and Barksdale Air Force Base, Louisiana, included the errors made by support personnel in mistaking nuclear weapons for unarmed missiles to be destroyed (despite distinctive markings) and the failure of a series of support personnel and flight-crew members to *fully* check what weapons were being hung on a B-52 for transport and what weapons were on an arriving B-52. Ultimately, the responsibility lies with those who designed the system and those who failed to monitor the detailed performance of the redundant system and the many partial failures that must have preceded this major lapse.

2. *Software problems at the "front end" are a special system issue.* Those working to reduce safety and security risks to a minimum are typically associated with the weapon system program, such as the CTM program, but the possibility exists—at least in principle—for errors to occur between the time of decision, its interpretation, translation, encoding, and transmission, and the time at which the message to execute is received on the weapon system platform. Especially in a fast-track system such as the CTM is intended to be, the possibility always exists for software errors that are not detected early. A program to ensure safety, then, must include the full end-to-end process, including aspects of command and control over which the program office has no control.

3. *Errors are often correlated rather than independent.* As illustrated by the recent B-52 fiasco, error or complacency by one person or one group of people can be "passed on," so that a sequence of intended checks turns out to be an initial failure to check followed by a sequence of pro forma nods. That is, what should have been sequential independent checks are instead correlated. In addition, of course, a system of checks and controls may fail if it has been poorly designed, so that the failure of a particular critical component means the failure of the whole. A familiar example is that of a building or area secured with formidable checkpoints with well-disciplined guards, but to which a construction crew is allowed easy entry by merely flashing a generic badge. In military systems, a parallel example might involve a critical command-and-control node.

4. *Errors of procedure are more likely to occur when equipment needed for different purposes is co-mingled and when people have more than one mission.* All else being equal, it is unwise for mechanics to have similarly shaped parts for different devices on their immediate worktable; for soldiers to be carrying weapons and ammunition that would be prohibited for their current mission; or, more to the point, for conventional and nuclear weapons to be co-mingled in ways that would make it relatively easy for the wrong weapon to be used—whether accidentally, in anger, or as the result of conspiracy.

5. *Security designs that include a combination of technical and procedural safeguards can be much more effective than either class separately.* The recent B-52 event occurred because of a sequence of human failures. It might have been readily avoided if, for example, an electronic system refused to permit mounting of the missiles because of a test indicating that the missiles were nuclear-armed. An electronic system might have refused to permit nuclear-armed missiles from being removed from the warehouse without a special electronic key available only to logistics personnel recently certified for such activities. However, computers and electronic systems are by no means foolproof and by no means do they always work properly. The classes of error that beset them, however, are different from those that can defeat processes dependent on consistent human adherence to procedures.

Planning for the CTM

The reason for including this discussion of safety is that the committee believes that those expressing worries about the safety and security of strategic weapons such those for CPGS are quite right to have concerns, but that solutions to the problem should come from careful system engineering rather than simple heuristics and impressions. That the B-52 incident represents an "unimaginable" lapse in the handling of nuclear weapons does not mean that all systems for handling, controlling, and using nuclear weapons (or strategic conventional weapons) are equally likely to have such mishaps. Different processes and protections are used for different weapon systems (e.g., B-2s, B-52s, ballistic missiles); some are better than others, and the relative goodness (reliability and effectiveness) can be assessed by independent reviewers with in-depth knowledge.

Recognizing the potential for error or loss of control of security or safety, the committee was very impressed by the preventive measures contained in the Navy's plan for the CTM. The committee was briefed in detail on the proposed missile and shipboard modifications and the operational procedures that would provide CTM command-and-control and nuclear surety for SSBNs deployed with both nuclear and conventional Trident missiles. Among these measures are separate safes for the firing keys of the nuclear-armed and the conventionally armed missiles (one safe would be held by the commanding officer, the other by the executive officer), incompatible electrical interfaces, two physical blocking elements in the firing circuit in the missile, software interlocks, physical examinations on loading, and unique missile designators. Many standard SSBN nuclear surety measures apply.

Standard nuclear surety procedures in system development will also mitigate the risk of an accidental nuclear launch in a CTM engagement. Personnel from the Naval Surface Warfare Center, Dahlgren Division, Virginia, who write the Trident fire-control software are closely monitored in a Personnel Reliability Program. Dahlgren personnel independently verify missile software developed by Lockheed Martin and vice versa. OSD periodically conducts an extensive 3-week review for all system changes.

CTM-equipped Trident submarines would not be the first nuclear-capable mixed-load platforms. In the 1980s, attack submarines carried nuclear-armed land-attack Tomahawk cruise missiles in addition to conventionally armed antiship Tomahawks. Earlier in the Cold War, smaller nuclear weapons for antisubmarine warfare (antisubmarine rockets and submarine rockets) and antiair warfare (one of the family of Standard Missiles) were carried together with conventional weapons on surface ships and attack submarines. Furthermore, SSBNs routinely operate with mixed loads during fleet commander's evaluation tests, during which nuclear payloads are switched out with test payloads before beginning a strategic patrol, and these reconfigured missiles are fired during the patrol under operational conditions.

In the committee's opinion, the Navy's Strategic Systems Programs (SSP) presentations to the committee address the nuclear surety issue well. It is clear that SSP recognizes the additional risks related to CTM and is exercising due diligence in its plans to mitigate those risks. SSP's track record gives one confidence. However, owing to the magnitude of the consequence of an accidental launch and the small amount of time that commanders in the field might sometimes have in prompt global strike engagements, the committee recommends the following: in addition to the risk mitigation measures proposed by SSP, OSD should appoint a red team to review thoroughly SSP's plans, searching specifically for ways in which, perhaps due to a series of concurrent failures, an accidental launch of a nuclear weapon might occur in an intended CTM engagement. SSP has used aggressive red teams in the past with excellent results.

With the above proviso, the committee concludes that the risk of an accidental launch of a nuclear weapon in a CTM engagement can be made very low, and it believes that the risk of an accidental launch should not by itself be a reason to decide against developing the CTM.

FINDINGS AND RECOMMENDATIONS

The committee reached the following conclusions:

- *Importance.* The CPGS mission is important and worthy of near-term priority and action. Indeed, extremely negative public reaction would be understandable in a few years if the DOD were *unable* to accomplish crucial conventional strikes (e.g., against terrorist leaders or terrorists moving weapons of mass destruction, or against a rogue state that continued to fire missiles at an allied capital for a period of hours) because of U.S. failure to have provided the option.
- *Near-term options.* The only credible near-term (2 to 3 year) option is the Conventional Trident Modification and (with an additional year or 2 delay) a variant proposed by the committee for consideration, referred to as CTM-2. The Minuteman option, which might have been expected to be a contender for the near term, turned out not to be as attractive as might have been expected (see Appendix I).
- *Longer-term options.* For the mid-term (e.g., around 2015), at least one additional option is available, the SLGSM launched from SSGNs. Thereafter, land-based options (CSM-1 and CSM-2) are proposed, with boost-glide capability if high-risk R&D proves successful over the next few years, as well as hypersonic cruise-missile options. These options are best seen as potential long-term options worthy of near-term R&D, but with IOCs not to be expected before about 2020.
- *The CTM.* The near-term CTM would be effective for an important class of missions, it would be marginal for others, and ineffective for others. Describing the CTM as a niche weapon is appropriate. The committee emphasizes, however, that a niche weapon may be highly valuable if it provides operational capability for addressing critical situations, which would be the case with respect to CTM for several of the scenarios considered. The technical means for ensuring the nuclear safety of a CTM system ensure that the risk of an accidental launch of a nuclear weapon in a CTM engagement can be made very low. For the reasons explained in Chapter 3, the committee believes that it is highly unlikely that a launch of a conventionally armed Trident would be misinterpreted as a nuclear attack; that the chances that such a launch would trigger an immediate nuclear response is very low and could be reduced even further by means of established cooperative measures; and that the "ambiguity" issue is not a reason to forgo the capability that CTM would afford.

- *An alternative.* The CTM-2 could be an immediate follow-on to the CTM, or possibly an alternative, albeit one that would add to delay in IOC and would have to build heavily on the development and testing of the CTM, as would the SLGSM. The CTM-2 would have greater payload volume than the CTM and, being a two-stage missile, under some circumstances could be distinguishable from the three-stage CTM to observers of the CTM-2 launch and trajectory.
- *Enablers.* A number of enabling activities are essential to mission success for CPGS. These activities, such as the precise determination of aimpoints, will sometimes have been done in advance and if so will not be a limiting factor, but they could also need to be done in a matter of an hour or so. Regardless of how quickly they must be accomplished, the quality of the enabling activities is critical, and the demands on those activities will grow over time if the ambition is to be able to strike moving targets. Major efforts are needed with respect to the enabling activities, not just the CPGS platform and weapon system.
- *Avoiding excessive constraints on R&D.* A congressional prohibition on CPGS R&D building on the Trident missile would have serious adverse consequences for almost any CPGS option, causing delays of at least 4 or 5 years. If Congress should allow Trident-related R&D but not authorize the CTM's deployment, the CTM-2 or SLGSM availabilities would not be much affected—perhaps by a year or so.
- *Special issues.* The committee recommends the following: (1) Congress should require that all CPGS options be developed with concepts of operations meeting very high, strategic, levels of safety and security—comparable to or better than the concept for the CTM, and planning for accomplishing this should include a review of command-and-control software, not just of weapon system issues, as well as rigorous red teaming to detect problems; (2) OSD should appoint a red team to thoroughly review the Navy's plans, searching specifically for ways in which, perhaps due to a series of concurrent failures, an accidental launch of a nuclear weapon might occur in an intended CTM engagement; (3) options for increasing the number of warheads per CTM, giving the CTM increased capability against terrain-protected targets, and perhaps improving the lethality of the flechette-based weapon used on the CTM (and CTM-2), should be examined immediately, since such options are potentially important and may be feasible even in the initial deployment; and (4) Congress and the DOD should resolve the pure-load versus mixed-load issue on its merits, adjusting the various options accordingly so that they are measured by the same assumptions.

3

Political, International, Policy, and Doctrinal Issues

The basic policy conclusion of the Committee on Conventional Prompt Global Strike Capability, based on its collective knowledge and judgment, is that there would be important political and strategic advantages to the United States in being able to strike high-value targets having time-sensitive urgency that could not be effectively engaged by currently available conventional strike systems. Existing conventional capabilities are extensive (and should be improved, where possible, to enhance capabilities in the interim before any conventional prompt global strike [CPGS] system can be operational), and no CPGS system could fill all the gaps in the U.S. ability to strike potentially important targets. However, CPGS would fill some gaps and add more options for dealing with threats to the United States and to its allies and to others whose interests are important to its own. Having those options available could be valuable in the campaigns against terrorism and against the proliferation of weapons of mass destruction (WMD) capabilities, as well as for larger-scale military efforts. (The nature of these targets and the proposed capabilities and limitations of the Conventional Trident Modification [CTM] and various other CPGS concepts are addressed in detail in Chapter 2.)

CPGS systems, however, present challenges and questions as well as opportunities. The essential policy judgment that must be made in deciding on acquiring a system—and on the doctrine and principles for its potential use—is whether the additional capabilities that CPGS would provide outweigh the costs and concerns that CPGS may present. The answer may depend on the type of CPGS system under consideration. On the one hand, for CTM, some challenges are smaller than would be the case with other CPGS systems because of the limited scale of the CTM program, CTM's relatively low cost and technological risk, and the opportunity to take advantage of the mature Trident system. On the other hand, some alternative concepts could be designed and deployed—with years of delay, perhaps many years—so as to mitigate some concerns that CTM would raise simply because it does build so directly on an existing strategic-nuclear system and has some limitations on its effectiveness.

THE NEED TO CONSIDER THE FULL RANGE OF ISSUES PRESENTED BY CPGS CONCEPTS

With the introduction of any significant new military capability, the doctrine for use of the system must be clearly defined and agreed on by policy makers and military commanders and planners. This is especially the case with a system such as CTM (or any other CPGS system): these systems represent a unique military capability, and their use (and perhaps even possession) has implications across the spectrum of diplomatic, policy, and military considerations that go far beyond even the complex and important technical and operational military questions presented. Setting the frames of reference for thinking through and planning for these issues is a critical step in deciding whether to develop and acquire the capability—and in preparing for its wise and effective addition to the nation's instruments of military power. Developing doctrine is an essential part of acquiring a CPGS capability, and it requires thinking through not just potential scenarios for crisis and conflict in which CPGS might be used, but also all of the implications of possessing such a capability.

The Primary Issues

The primary issues that have been raised with respect to CPGS are these: (1) some CPGS systems might raise ambiguity problems because of using strategic-nuclear delivery systems (notably, the Trident submarine-launched ballistic missile [SLBM]); (2) the advertised promptness of the CPGS capability will not be achievable for years because it requires exquisite intelligence, detailed surveillance, rapid decision making, and so on; and (3) the CTM concept would degrade the quality and certainty of the nuclear deterrent force. The committee analyzed all of these issues in considerable detail. Additional matters addressed in this chapter include command and control and the requirement for presidential authorization, the potential for inappropriate or mistaken use, the prevention of an accidental launch of a nuclear weapon when a conventional strike has been ordered, overflight and debris, access to forward basing, proliferation, arms control and treaty issues, and strategic considerations.

The consensus of the committee that emerged after study was that (1) the ambiguity problems have been overstated and can be substantially mitigated; (2) there are likely to be cases where intelligence could be available and decisions could be made promptly enough to make a short execution time highly valuable, even though minimizing other parts of the end-to-end planning time will take years to achieve (as discussed in Chapter 2 and not repeated in this chapter); and (3) any negative effects on the nuclear deterrent force would be quite small. The remainder of the chapter addresses the broader range of issues systematically, but the complexity of that discussion should not obscure these bottom-line conclusions.

A Broad View of the Issues to Be Considered

Decisions on CPGS entail consideration not only of technological possibilities and military missions, but also of the challenges and opportunities that having (or lacking) such a system would present for national policy and national interests. Moreover, in deciding whether to acquire such a system, it is essential to take into account the doctrine and procedures that would be developed and applied to determine how and when the options afforded by CPGS should be used (and withheld). The possession of CTM, or of any other CPGS system, would add to the nation's options—that is, to what the military is technically capable of doing. However, it is essential to have for CPGS, or any other system, a structure for operating it and making judgments about what it is to do—and consideration of those issues must be part of the decision on whether to acquire CPGS. The United States has for many years had long-range strike capability, which it has also exercised, as illustrated by the 1986 strike on Libya and antiterrorist-related strikes into Afghanistan and Sudan in 1998. Both involved aircraft. Strikes might be conducted tomorrow with such systems if the location of terrorist leaders were known. It might seem, then, that most of the policy-level issues have long since been resolved and precedents established. Arguably, however, conventional prompt global strike raises some new issues.

The specific questions that need to be considered in developing policy and procedures in the CPGS context include the following:

- What are classes of high-value targets that the United States may want to strike?
- What intelligence will be needed, and how fast can it be provided and exploited?
- What information can be gathered and processed in advance (e.g., early warning to focus preparations, precise geolocation of plausible targets, and preliminary or contingent decision making)?
- What is the decision-making process? What is the chain of command?
- What will have to be done to plan the mission? What preplanning can be done in advance of facing a decision on actual execution? When, for example, can a contingent decision be made—to use the system if certain information is available about an identified type of target, subject only to a rapid, final confirmation by the President?

- Given sufficient warning-and-preparation time, how effective and reliable will the CPGS be?
- Apart from the nature of the target, what information will be required as essential elements for making a decision—for example, probability of success (taking into account possible countermeasures and weapon system constraints), likely collateral damage (including the possibility of fallout from destruction of a WMD facility), overflight and debris impact, subsequent opportunities for attack, time to execute the strike after the order is given, and so on?
- What are the alternative ways of attacking the target?
- What consultations or notifications should be made in the United States (e.g., Congress) and in other countries?
- What are the possible and likely reactions of the target country and other countries? What is the next move for the United States in the aftermath of success or of failure?
- Is there any chance that an attack with a CPGS system will be misinterpreted as a nuclear attack, particularly in the midst of a crisis?
- What are the longer-term reactions of other countries to the existence of the capability (and the use) of a CPGS system? What would be the effects on proliferation?
- What are the vulnerabilities and limitations of the system to be deployed, and what should be done to correct or mitigate them?

These issues are not fundamentally different from those that would arise with the use of any military force, but CPGS gives the capability to act very quickly, and therefore it is all the more critical that these questions—and others—be thought through and prepared for in advance. The alternative is errors arising from failure to consider all the relevant issues, or decision-making and processing delays that lose the advantage of having the technical capability to act with sufficient speed.

Faster execution time will by definition make it possible to strike faster. Whether this gain will be critical is entirely dependent on the particular situation and its circumstances. However, the interval from launch to impact on the target is far from the only factor in determining whether the United States has a capability for prompt global strike. Promptness of strike is measured not only by the time interval between launch and impact but by the whole end-to-end time line. There must still be a careful and informed decision to use the capability, and the attack must be properly planned. Intelligence must be gathered, distributed, and evaluated; decision makers must be informed of the intelligence and the options and consider the pros and cons of ordering an attack; detailed operational plans, including precise geolocation of the target, must be prepared, approved, and transmitted; and necessary strike preparations, including bringing weapons to ready-to-launch status, must be made.

Some of these necessary antecedent steps can be done in parallel and some can be done well in advance. This is particularly true if the need or opportunity arises out of an ongoing confrontation in which contingent decisions can be made to use (or not to use) the CPGS capability if the opportunity presents itself, subject only to a final confirmation by the President. However, it will always be the case that these steps take time and that they require enabling capabilities, for example, for intelligence gathering, target location, and mission planning.

If the United States is going to deploy a CPGS system, all of these issues must be addressed. This is so because being able to gain the full benefit of an investment in a prompt execution capability requires both that enablers supporting that capability be available and that steps be taken to ensure that each element in the end-to-end time line—not just the time from launch to impact—is made as short as it can be. If it is valuable to invest in a system that has the advantage of a very short time of flight, then it is also valuable to minimize the time required to use the system effectively, bearing in mind that there must be time for deliberation and for ensuring that the strike is properly planned and that hair-trigger decisions are likely to lead to errors.

Moreover, shortening the execution time once an order reaches the launcher does not necessarily make the overall time to decide on and plan an attack shorter; nor should it always do so. There may be cases in which the principal benefit of a shorter execution time is that it gives more time for other steps in the process to be carried out properly, while still not missing the opportunity if the decision is made to strike.

Planning for making effective and prudent use of the execution capability also requires addressing doctrinal issues and decision procedures as well as technical parameters. Accordingly, a key part of the decision to proceed with CTM, or any other CPGS system, should be a commitment to a major comprehensive effort to develop theory, doctrine, and commonly accepted understandings (analogous to that for nuclear theory, including extended deterrence) as well as to taking other steps necessary to minimize the total end-to-end time line.

COMMAND AND CONTROL AND THE REQUIREMENT FOR PRESIDENTIAL AUTHORIZATION

Although CTM's absolute firepower is very limited compared with that of many other modern military systems, its use would almost certainly occur in highly sensitive political contexts, because of either the nature or the location of the target. Almost by definition, potential targets of CTM would be highly important to national interests (e.g., terrorist leaders, WMD supplies or facilities), but executing the strikes could have ramifications well beyond the immediate operation. Opportunities to use CPGS would often raise policy, legal, and diplomatic issues regarding the use of American military force against targets located in the sovereign territory of nations where the United States was not simultaneously conducting other military operations. These issues have arisen often in the past, but the fact that CPGS systems offer the opportunity to reach such targets rapidly will not reduce the significance of the issues. CPGS would in some situations provide a military option that otherwise would not exist, but the availability of that option should not blind decision makers to the existence of alternatives, both military and otherwise.

These aspects of an attack using a CPGS system mean, in the committee's judgment, that a basic element of the doctrine for CTM, or any other CPGS system, is that the authority for the execution of these very limited, high-value strike options should rest exclusively with the national command authority, that is, the President acting through the Secretary of Defense in transmitting orders to the military, even in the midst of a conventional conflict (such as in Iraq or Afghanistan currently). Such a reservation of execution authority provides maximum assurance that all of the relevant political and strategic considerations relating to the strike being proposed (e.g., ambiguity, likely immediate and longer-term responses, collateral damage, consultations, alternative strike options, and so on) are taken into account. Presidential authority could, in appropriate circumstances, include instructions to the relevant military commander to employ the system if conditions defined by the President were met. This "escape clause" should be used with extreme caution, however, and except in circumstances that are both truly extraordinary and very clear-cut, should always be subject to last-minute, prearranged confirmation by the President.

In setting up a system for the authorization of strikes using CTM, it will be essential to establish a clear chain of command below the President. Presumably, the relevant regional combatant commander would have operational control of strikes against targets in his area of responsibility (AOR), as is the case today when long-range strike aircraft based outside a regional commander's AOR execute strikes at targets in "his" region. However, whatever the details of the chain of command, CPGS systems (like nuclear weapons) should not be considered simply new capabilities to be available to field commanders to employ on their own authority.

POTENTIAL FOR INAPPROPRIATE OR MISTAKEN USE

A concern has sometimes been raised that CPGS capability should not be acquired because it could be misused. More specifically, those who hold this view argue that CPGS would give the President a non-nuclear option that might be particularly susceptible to misuse, because decisions to use the system could seem "too easy": that is, the capability would facilitate military attacks deep in the territory of other countries, not just without using nuclear weapons but without deep and extensive commitment of other elements of the American military and other resources, and with little immediate risk to American personnel.

The defect in this argument is that it applies too broadly. Any military system could theoretically be used incautiously, but the United States cannot, for reasons that go far beyond CPGS, make defense-acquisition decisions on the presumption that presidents will act foolishly. Developing a sound procedure for decision making will help ensure that the mere capability for very rapid action does not become a pretext for ill-considered strikes. The potential for inappropriate decisions—which exists with any military capability—should be dealt with by careful planning and procedures for using a new system, not by denying the nation the possibility of using the military capability wisely. The committee believes that in the emerging doctrine, the use of CPGS systems should be so much an exceptional, rather than a routine, operation as to make that casual, ill-considered decision making very unlikely. Moreover, it is untenable to say that the country should forgo having the means to strike effectively against targets threatening its national interests. There is some chance that capabilities could be used unwisely; the relevant question is how to minimize the risk of "unwise use."

There is a somewhat separate concern that the fleeting nature of the target will produce pressure for rapid decisions and result in ill-considered action that might be avoided if more time had been taken to reach a decision. However, the pressure for very rapid decision on CPGS operations would be driven by the nature of the target, not the speed with which the attack, once ordered, can be carried out. Potential CPGS targets are of sufficient importance that if CPGS were unavailable, those targets might well be struck, albeit less effectively, with other conventional systems. In that case the time for decision would be even more limited than with CPGS, and the effectiveness would, almost by definition, be less. Indeed, there are cases in which the greater speed of execution would allow *more* time for deliberation and reduce the perceived need for early delegation of authority to local or even tactical commanders.

The committee's judgment is not that these arguments concerning misuse or error are entirely irrelevant or absolutely wrong. The fact that a CPGS system could by its inherent nature be used rapidly, simply makes it all the more important that the decision to use it be made carefully including, where possible, deciding in advance for (or against) its use in the event that an opportunity arises. The committee believes that the increased possibility of successful uses of CPGS in cases where that is the right decision justify having the capability and outweigh the theoretical possibility of inappropriate uses. Moreover, while it is obviously important to do all that can reasonably be done to avoid mistakes, the fact that CTM attacks would necessarily involve limited numbers and limited effects (compared with options involving existing conventional forces that might require defense suppression, a larger attacking force, and risk to personnel, not to mention comparison with nuclear attacks) would reduce but not eliminate the potential ramifications of mistakes.

IMPACT ON NUCLEAR DETERRENCE AND STABILITY

At a minimum, it must be a basic premise of CTM, or any other CPGS system, that it should not weaken nuclear deterrence. In particular, introducing CTM as a conventional system on the Trident nuclear-powered ballistic missile submarines (SSBNs) should not be permitted to interfere with core nuclear deterrence or with the readiness, deployments, training, discipline, day-to-day operations, research, and in general the focus and resources necessary to maintain that core deterrence. The concern is sometimes expressed that because CTM adds a conventional mission to a platform and support system hitherto exclusively devoted to the nuclear deterrence mission, it will risk a lack of focus and diversion of attention and priority (including funding for technological work needed to keep the force fully capable of meeting its mission) from strictly nuclear issues.

As for a potential lack of sufficient weapons on SSBNs in appropriate patrol areas to maintain necessary levels of available forces to support nuclear war plans, the very modest reduction in on-station nuclear loads and the adjustment of patrol areas necessary for the CTM mission would not compromise the nuclear mission. (Because the United States has reduced the number of reentry vehicles [RVs] on Tridents to below their design potential, it is the committee's judgment that the off-loads for the two CTM missiles per on-patrol SSBN could be offset by increasing the number of RVs on other missiles on the same submarines.) For reasons explained below, the dangers flowing from a possible identification of the location of the submarine resulting from the launch of a CTM are minimal.

Two concerns that have been expressed are that the patrol areas would no longer be optimized for survival and that the launch of a conventional Trident missile would reveal the SSBN's location, thereby endangering its survivability for the nuclear mission. Although it is true that the patrol areas would not be "optimal," they would still be extremely large and beyond what is considered needed by any current analysis. No other nation possesses the assets (nuclear-powered attack submarines, patrol aircraft, and so on) to conduct large-area search-and-destroy operations against U.S. SSBNs. As for launch revealing location, that problem has been studied and dealt with over the years. The problem would be no worse than with a nuclear warhead, and the Navy long ago developed techniques to protect SSBNs after missile launch. These techniques include deceptive trajectories to complicate backtracking, and rapid SSBN movement from the launch point. Within a short period of time, an SSBN's location would be unknown within an area large enough to deny plausible effective attack. Finally, it should be noted that in the event that circumstances changed, with the emergence of a serious threat to U.S. deterrent forces, SSBNs could readily revert to patrol areas and practices fully optimal for nuclear deterrence.

With respect to the "softer" elements of maintaining the nuclear mission, the committee concludes that the Navy has developed the plan for CTM in a way that will hold the conventional role to the same high standards as for nuclear missions and will maintain attention and priority on the nuclear mission. The very fact that—unlike other systems with a potential nuclear but a primary conventional role—CTM would necessarily be employed only infrequently and on very special missions means that there is little potential for a routinized conventional mission to drive the nuclear role into the background; there is little risk that nuclear missions would come to be seen as secondary to the "real" conventional mission, with an attendant reduction in attention to all the no-doubt tiresome and petty-seeming rules and procedures that (properly) go with nuclear weapons.

Some argue that having an extensive CPGS capability would make it possible for the United States to declare a policy of no-first-use of nuclear weapons—that is, limiting U.S. use of nuclear weapons to a response to an actual nuclear attack. Apart from the more general arguments about whether a policy of no-first-use is wise, it is difficult to believe that the capacity for a very limited number of quick, highly accurate conventional attacks is significant to the debate. For example, an effective non-nuclear response to a massive chemical or biological attack, whatever its wisdom or feasibility, would hardly rely greatly on the capacity of CPGS systems for rapid and precise attacks. Rather, the response would rely primarily on overwhelming force. CPGS is no easy—or tempting—road to no first use.

From a very different perspective, it is sometimes argued that a reliance on CPGS weakens the deterrence of WMD attack (or other equally grave provocations short of a nuclear attack), because investment in CTM would imply that the United States does not intend to use nuclear weapons in such circumstances: that is, that CTM betrays a lack of will in a context in which the nation would very much want adversaries to believe that it would use nuclear weapons. This concern seems overblown. As noted above, neither CTM (nor even other, more extensive CPGS systems) would likely be critical one way or the other to a non-nuclear response to a major WMD attack. More important, having a conventional option for certain limited missions does not in any way reduce the credibility of threats to respond with nuclear weapons to grave provocations.

Quite apart from relatively technical concerns with ensuring that CTM does not reduce the effectiveness of the nuclear deterrent (and the important "ambiguity" question, discussed below), issues have been raised concerning the potential impact of CTM on nuclear stability. It could be argued that CTM or other CPGS systems could make nuclear war more likely because they would enhance U.S. capabilities in ways that render more likely U.S. attacks on high-value targets in nuclear-armed countries, to which those countries might respond by using nuclear weapons. Of course, any attack on a nuclear-armed country is a very serious matter, simply because of the possibility of escalation, but the probability of a nuclear response to such a conventional attack is surely lower than the probability of a nuclear response to a nuclear attack.

CTM and other CPGS systems are sometimes argued to have a more general effect of raising the nuclear threshold by allowing conventional attack where nuclear weapons would otherwise have been used. The degree to which the availability of an additional conventional option raises the nuclear threshold for the United States depends on the degree to which a strike with CTM would actually substitute for a nuclear strike, as contrasted to (merely!) giving the President an effective option in place of a nuclear alternative or inaction in a situation where there would otherwise have been no effective U.S. action.

Opinions differ on how often, if ever, a U.S. President would actually use a nuclear weapon in the sort of scenarios for which a CPGS system would be designed. It is certainly possible to conceive of cases where a threat is so great that there is a very real possibility that the United States would use a nuclear weapon (e.g., in a case of absolute confirmation of an impending nuclear attack on a U.S. city), and a CPGS capability might allow the President to substitute its use for a nuclear strike and thereby act effectively while avoiding a nuclear strike. In this very limited sense, CPGS would raise the nuclear threshold. (There is a separate question about whether in a case where the President would actually have ordered a nuclear attack, any CPGS option would appear sufficiently effective—and dramatic—to be an adequate substitute.)

More important and certainly more likely, CPGS would be a valuable new instrument of national policy because in extremely serious cases it would avoid the dilemma of having to choose between using a nuclear weapon or making no response at all. The committee believes that in the face of such a dilemma, the disadvantages of using nuclear weapons are such that there are very few cases in which the President would actually choose to use a nuclear weapon. But without CTM or some other CPGS system, the lack of an effective and sufficiently prompt non-nuclear response and the appropriately extreme reluctance to use nuclear weapons for any but extraordinarily grave threats would leave the United States in the position of being unable to take any effective military action.

On balance, it is the view of the committee that CPGS would have little effect on nuclear stability or deterrence because in most cases it is essentially an improvement—perhaps a decisive improvement—over existing but less satisfactory conventional options (or for no response at all). In virtually all cases in which CPGS might be used, even in those cases where other conventional options would be ineffective, there is little, if any, real possibility of the United States actually using a nuclear weapon. However, providing an option to fill part of the gap between what current conventional systems can do and the point at which a nuclear attack would be a wise action is an important advantage of CTM, whether that gap is large or small.

PREVENTING AN ACCIDENTAL LAUNCH OF A NUCLEAR WEAPON WHEN A CONVENTIONAL STRIKE HAS BEEN ORDERED

Any CPGS concept must ensure that there is no possibility that an order to carry out an attack with the CPGS system would accidentally result in a nuclear attack, or that such an order would accidentally result in the launch of a missile with a nuclear warhead even if no nuclear detonation occurred because the weapon did not arm. This is a particular issue in the case of CTM, because the system involves having conventionally armed missiles on a submarine that also carries nuclear weapons (see Chapter 2 for a discussion of nuclear safety issues).

There are multiple measures—effectively "fail-safe" procedures and mechanisms—that can reduce, and in principle eliminate, any risk of an accidental launch of a nuclear weapon when a conventional strike has been ordered. These include taking procedural and physical steps to prevent (and if possible make physically impossible) (1) the launch of a nuclear missile in response to a conventional launch order, (2) the loading of nuclear-armed missiles into launch tubes for conventionally armed missiles, or (3) the transmission of a nuclear launch order when a conventional launch order is intended.

The Navy's plans for CTM are designed to meet these objectives by, among other things, including distinct command-and-control systems for the two types of missiles on the submarine (details are provided in Chapter 2). In addition, CTM would exploit an overall command and operational management system that has been repeatedly used to launch without error training missiles from an SLBM also armed with nuclear missiles. These "fail-safe" arrangements should combine procedural steps and physical controls—that is, procedural steps relying on the following: the careful selection of personnel, application of a screening program to personnel involved with CTM, well-understood and well-designed processes (e.g., command-and-control protocols), continual training and exercises, and similar personnel-oriented measures; and physical controls, such as air gaps between conventional and nuclear launch systems, and required external inputs to arm missiles and warheads. These measures would mean that errors, if they occur, do not result in an accidental launch or detonation of nuclear weapons.

NUCLEAR AMBIGUITY

The most prominent objection on policy grounds to the conventionally armed Trident missile, and to some degree to all CPGS proposals, has been the "nuclear ambiguity" concern: the possibility, with potentially catastrophic consequences, that a launch could be misinterpreted by foreign observers as a nuclear attack. The concern most commonly raised is that a major nuclear power that regards itself as a potential adversary of the United States, that is, Russia or China—having detected the launch of a CTM from a deployed Trident SSBN and realizing that the missile was coming out of a ship that carries similar missiles with nuclear warheads—might misinterpret the nature of the event that it was observing.

Before going through the issue more systematically, it is noted that a key element of the concern as most frequently articulated is that *in the past*, SSBNs and Tridents have been associated exclusively with nuclear weapons. If today a Trident were launched in the direction of Russia or China from an ocean area where the United States did not routinely do missile tests, and if that launch were detected, there would be little alternative to interpreting the missile as being nuclear-armed. (Even in that situation, the apparently targeted country would have every reason to refrain from an immediate nuclear response and every reason to doubt that the United States was actually starting a nuclear war with a single missile.)

One thing, and perhaps only one, is absolutely clear about the ambiguity problem: Simply using something other than Trident missiles (or another legacy ballistic system) as the delivery platform does not avoid the problem entirely. Indeed, ballistic missiles, whatever their nominal maximum range, have the virtue that in the event of an actual launch, well-understood and extensively practiced surveillance systems can quickly acquire sufficient data to determine their trajectory and hence the observer can be highly confident of their target. There have been hundreds of launches of long-range ballistic missile rockets for testing, training, space launch, research, and other purposes over the past several decades, and these have served to familiarize warning system operators with the identification of the nonthreatening nature of such events. If the maximum range of a new CPGS ballistic missile system (e.g., a Submarine-Launched Global Strike Missile [SLGSM] or a two-stage Trident based in the Indian Ocean) is too short to pose a threat to critical targets inside Russia or China, this would decrease the ambiguity problem. However, detection systems may not be able to identify what particular type of missile has been launched, and there may be uncertainties about the launched missile's maximum potential range.

Nor is prior association with nuclear weapons—that is, being a legacy system—a major factor, much less a decisive factor contributing to the misinterpretation of a conventional attack. There simply is no "bright line" between nuclear and conventional systems when relatively long-range platforms are being considered. While long-range ballistic missiles are now associated exclusively with nuclear weapons, that was once equally true of many systems that have since come to have extensive conventional roles (e.g., the B-52, B-1, B-2, fighter bombers, sea-launched cruise missiles [SLCMs], air-launched cruise missiles [ALCMs]). If the United States develops and deploys CTM (or any CPGS system derived from an intercontinental ballistic missile [ICBM] or SLBM), even the psychological association of the system with an exclusively nuclear role will erode. If CTM itself were developed and deployed and then finally used—after years of discussion, testing, exercises, exchanges, and so on—first impressions and intuition would be different: A conventional attack would probably be the first and dominant "explanation" of what was going on (as it is today when the United States uses in conventional combat systems what were once exclusively or primarily nuclear—notably, long-range bombers or cruise missiles).

In the case both of other CPGS systems that might be developed in the future to have operational advantages over CTM and of other ballistic-missile-derived systems of far greater maneuverability (e.g., boost-glide missiles, aircraft), it will be impossible for other countries to be certain of the target until after it has been hit. Merely avoiding a declared nuclear role for a new and highly capable CPGS system (or relying on a system with no nuclear legacy) would not eliminate the ambiguity problem, or even necessarily reduce it, relative to CTM or other legacy-based CPGS systems. Any system—old or new—that the United States might designate as conventional-only (e.g., a submarine with an exclusively CTM load, or a new missile or hypersonic glide vehicle) could be armed with nuclear weapons. Other countries could never be confident that there is not a nuclear warhead on a new U.S. system simply because the United States says so. It is true that the United States has never put nuclear weapons on a delivery platform without testing and exercising it and its attendant nuclear security and command-and-control system in that role (though without an actual nuclear weapon); however, other countries would not necessarily regard that as decisive, particularly insofar as the issue is the use of such a platform in a limited but highly effective role as a precursor to a larger attack. Even with an entirely new CPGS system, it would be necessary to take measures to address the ambiguity problem.

Having started with this observation in the belief that it is telling, the committee now addresses the issue more systematically and, admittedly, with some complexity. There are two, logically and practically distinct, aspects of the ambiguity issue. The first is the possible misinterpretation by an observing nation of a conventional CTM strike on a third party as a nuclear strike on its own territory. The second is a mistaken conclusion by an observing nation that a conventional CTM attack headed toward its territory is a nuclear attack.

The issues to be considered in analyzing the ambiguity question are as follows:

- Who would be able to detect the launch?
- If a foreign nation were to detect the launch, would it be able to identify the missile type correctly and estimate the trajectory of the missile or of the reentry vehicles (RVs?)
- If the launch were detected by a foreign nation, what would happen?
- Would that nation's nuclear forces and surveillance systems be alerted and, if so, what would be the consequences?
- Would a "retaliatory" strike be ordered?
- Even if there were no immediate adverse effects, what long-term reactions might be triggered?

It is certainly appropriate to give serious attention to the ambiguity problem. The stakes are extremely high. In some sense the issue is not whether misinterpretation is unlikely, but rather, how unlikely is unlikely enough? In this connection, context matters greatly: reactions in time of general calm could be very different from those during a general crisis. Today and for the foreseeable future, U.S. relationships with Russia and China are less tense than in the Cold War era. In some important sense, there is less danger of accidental war, ambiguous launch interpretation, and so on. This is not to say that the technical issues to which this chapter now turns are unimportant, but only that the larger context against which they will be seen has changed in a way that lowers the greatest risks.

If a proposal were made for the use of CPGS against a third country at a time of increased U.S.-Russian tension, decision makers would need to take into account that in such circumstances there is a higher probability of overreaction or mistaken interpretations of events—of an inclination to put the worst possible face on events. Moreover, quality of information is an important factor. If a launch cannot be detected, there is no risk of overreaction; if detection and tracking capabilities are good, the same is true. Only detection with incomplete ability to understand the event (e.g., satellite detection of a launch without good radar-based capacity to predict trajectories) would be potentially dangerous. The dangerous condition is one in which the observer has imperfect information, as opposed to either no information at all or very accurate information.

The problem of "misinterpretation" of a CPGS attack on a nuclear-weapon state (i.e., the target country knows it is being attacked and may be unsure of the nature of the attack) is potentially of particular seriousness. If the target country has nuclear weapons of its own, an attack that is (correctly) perceived as directed at critical targets deep in its territory could trigger a nuclear response. It is far less likely, but not impossible, that a nuclear power, knowing that the United States had the capacity to mount CPGS attacks, might be motivated to adopt a policy of responding with a nuclear counterattack to any U.S. conventional attacks on critical targets.

The existence of these risks means that any use of CTM, or any other CPGS system, would have to be carefully considered in the broader context, particularly if the question were one of an attack on a nuclear power. Nonetheless, there are many reasons to regard the risks as sufficiently low and manageable that, while their existence needs to be taken into account—and the use of CPGS even forgone in certain cases because of concerns about escalation—they do not constitute a reason to forgo acquiring the capability. The risk of escalation makes any military action against a nuclear power a grave matter indeed, but these risks are no greater just because a CPGS system is involved.

Consider first the most likely case—a conventionally armed ballistic missile launch directed at a country other than Russia or China. Currently, there is no reason to believe that any foreign country other than Russia has a warning system that could detect such a launch. Assuming that Russia would detect the launch, the warning system would also likely have sufficient tracking to conclude in a few minutes that the target was not in Russian territory. There is strong historical evidence to support this conclusion. Because of flight tests and space launches, Russia, like the United States, is accustomed to detecting and monitoring missiles and rockets after launch to establish their trajectories and confirm that whatever they are, they are not attacks directed at their own country. Even if the Russians were unsure of the trajectories, they would certainly discern the necessarily very limited scale of a CPGS attack and would be unlikely to conclude that the United

States was starting a nuclear war with Russia in a "bolt-from-the-blue" attack with so few missiles. (Indeed, in discussing the much-publicized "Norwegian sounding rocket" incident, Russian spokespersons acknowledged recognizing that the event, though puzzling and even worrying, was not a U.S. attack for precisely this reason—its small scale.[1]) Correct interpretation would be overwhelmingly likely if, for a period of years, Russia had been reading about, talking to U.S. experts about, and observing exercises in the use of CTM or some other CPGS system. By the time the system was deployed and operational, the natural interpretation of its launch would be that, as had long been anticipated, the United States was exercising its CPGS capability.

It is likely that when their warning system detected any other than the most routine foreign missile launch, Russian commanders would increase the readiness level of their surveillance and command systems. That step alone does not increase the danger of a nuclear response, because it does not affect the alert state of the forces themselves—only of the command and surveillance systems. Indeed it is certainly intended to cause the situation to be monitored more closely and would cause greater attention to be paid to actions that might inadvertently cause escalation. Even if the Russian civilian leadership did mistakenly conclude initially that Russia was the target of a small nuclear attack, it would have every reason not to order an immediate counterattack because the few missiles in flight could not significantly degrade the country's ability to respond after the situation was clarified. This would be true even if Russia had a policy of "launch under attack" for major attacks or was concerned about the possibility of a precursor nuclear attack disabling its warning system.

Today other nuclear-weapons states (including China) appear to have very little capacity to detect a ballistic missile launch (and therefore very little potential for misinterpreting what they do not know). When China develops early-warning capabilities, as it presumably will in time, the observations above about the Russian capacity to analyze a ballistic missile attack correctly—and to refrain from nuclear response until it has—would apply. However, in the Chinese case, because China's nuclear forces are so much smaller than Russia's, an erroneous perception that a small nuclear attack was underway would perhaps be somewhat more likely to be regarded as a serious threat than would be the case for Russia. The word "perhaps" is used here because the possibility seems very small, and if China were concerned about bolt-from-the-blue attacks, it would have many mechanisms by which to improve the survivability of its forces.

What is at issue is the relative risk of an accidental nuclear war and of the consequences of not using a CPGS weapon. If CTM or another type of CPGS system is deployed, everything reasonably possible to further reduce the risks should be done, if only because there is always some risk that an unexpected event—the

[1]Geoffrey Forden, Pavel Podvig, and Theodore A. Postol. 2000. "False Alarm, Nuclear Danger," *IEEE Spectrum,* Vol. 37, No. 3, March.

sudden appearance of a unscheduled missile launch—might somehow set off a chain of unintended and supposedly impossible reactions. The stakes are so high that it is important, if any form of CPGS is deployed, to take additional measures to further increase the chances that Russia (and China) would not misread its use. These could include cooperative measures to increase information about the system and its operation (notification, transparency arrangements, a joint warning center, data exchanges, participation in and observation of exercises, inspection regimes) as well as extensive and candid discussions of the nature of the system and the U.S. doctrine for its use. There are also technical devices that could be installed on a mutually agreed basis that would provide additional confidence that a CPGS system declared as armed exclusively with conventional weapons had not been covertly nuclearized. An example of such a device would be the continuous visual monitoring system outlined in Appendix H of this report.

All the foregoing said, it is impossible to avoid the conclusion that whatever the risk of misinterpretation might be, it would be higher in cases where the issue was not a launch against a third country, but an actual attack on targets in Russia or China (albeit conventional and limited). Such an attack could, by definition, occur only at a time of very serious tensions and, quite possibly, active military engagement, between the United States and the target country. In such a context, the chances of overreaction, while still extremely small, would be higher than in other situations. While both sides would have an interest in controlling escalation, the combination of tension, suspicion, and uncertainly would be dangerous, and the risk of events getting out of hand would be increased.

This is not to say that potential use of CPGS even against a nuclear-weapons state should be ruled out in all circumstances. The problem of misinterpretation of a prompt precision attack on a nuclear-armed state is a special case of the general and very serious problem of the control of escalation in a conflict between nuclear-armed states. It is obvious that any such conflict carries great risks, but it could only come about if great interests were at stake, so that refraining from military action for fear of escalation could carry great costs. The challenge of using military force in a conflict with a nuclear power without unintended escalation is by no means limited to CTM (or any other form of CPGS capability), but it does mean that the use of a CPGS capability to attack highly critical targets deep within the territory of Russia or China would require great justification and consideration. (It is difficult to conceive of situations in which the benefits of using CPGS against targets in Russia or China would outweigh the risks.)

In sum, concerns about nuclear ambiguity will always be a potential issue, one whose significance depends not primarily on the technical characteristics of the CPGS system but on the context, scale, and target of the attack and on the degree to which transparency and confidence-building measures have been employed. There will be circumstances where ambiguity concerns might and should make a President less likely to use CPGS, especially in a crisis or for attacks against a nuclear power, but that is no reason to deny the possibility of its use in other situations where the risk of misinterpretation is not present. Moreover, the risk arising from the ambiguity of CPGS systems is one of the lesser risks—militarily and politically—associated with attacking another country, especial a nuclear-armed country.

OVERFLIGHT AND DEBRIS

As discussed in the section above on ambiguity, the overflight of third countries situated between the attacking and the targeted countries (or flight paths that generate concern about impending overflight) may lead decision makers to assume mistakenly that their own country is under attack. Such concerns are particularly important when the third party has both the ability to detect overflight and the ability to respond militarily. For CPGS systems that rely on air-breathing systems, overflight may also render the missile vulnerable to the air defenses of third parties.

In addition to issues of ambiguity and air defenses, overflight can pose political problems. Many countries would protest alleged violations of their airspace, as Japan did when North Korea conducted a test of a ballistic missile on a trajectory over Japan. Such protests could occur even if the third party had no capability to detect the overflight, because in most cases the fact that overflight had occurred would be obvious after the fact. Such protests could have significant political ramifications and would have to be taken into account in the decision to order a strike. Potential political problems could be mitigated by asking for overflight permission in cases where such permission is likely to be granted, or by notifying the government about the planned overflight in cases where permission would likely be denied, but in both cases unacceptable security compromises or delays in execution would be possible.

In most cases, the political consequences of overflight, apart from ambiguity concerns, would not be a decisive or even an important reason not to conduct a strike. The longstanding position of the United States is that transit through space on ballistic flight paths or orbits is not a violation of national sovereignty or airspace. Moreover, the United States has conducted nonconsensual overflight with aircraft when this was deemed necessary for very high priority operations. One example is the 1998 U.S. cruise-missile strikes on al-Qaeda training camps in Afghanistan, which involved overflight of Pakistan. CPGS strikes presumably would be rare, and limited to the most important and time-urgent targets, in which case possible post-strike protests of overflight of most third countries would not be a major factor in a decision to strike (although it might play some role in mission planning).

Thus, the ability of a particular CGPS system to avoid overflight of selected territory should be considered an advantage, but not a decisive one. Submarine-based systems, such as CTM and SLGSM, have the advantage that submarines can in some cases be positioned to avoid overflight of selected countries. The most important overflight concerns—those related to potential ambiguity problems involving Russia—can be addressed by designating routine patrol areas that avoid overflight. But the repositioning of submarines to avoid overflight of other countries could take many hours or even days, which might not be compatible with the prompt nature of the CGPS mission (and, in the case of CTM, would interfere with the core nuclear-deterrent mission of the submarines). Boost-glide and air-breathing systems would have substantial maneuver capability on their path to the target, which could be used to avoid overflight of some countries or air defenses. All else being equal, this is an advantage, but it is not dispositive.

Political concerns regarding overflight would be amplified by debris falling on a third country, which could result from the normal operation of the system (e.g., third-stage rocket motor) or from a system failure (e.g., the premature impact of a warhead or the crash of a cruise missile). Such concerns would be amplified further if the debris resulted in casualties or property damage. As in any use of military force, the potential for civilian casualties or other collateral damage would be a factor in the decision to order a strike. In most cases, the risk of collateral damage from overflight (as opposed to warhead impact in the intended target area) would be minimal, even for ballistic systems in which there can be no choice about the flight path from a particular launcher to the chosen target. In the case of CTM (and presumably SLGSM), the final stage of the missile contains enough propellant to allow mission planners to divert it hundreds of kilometers from its normal path, toward a body of water or a sparsely inhabited area, in order to minimize collateral damage. Boost-glide or air-breathing systems can take flight paths that avoid densely populated areas, but as with ballistic missiles, they are vulnerable to failures that could deposit debris anywhere along their flight path, which could diverge substantially from the planned flight path. CPGS systems could be equipped with systems to self-destruct automatically or to impact in uninhabited areas in the event of deviations from planned flight paths or of other failures that would prevent the warhead from reaching its intended target.[2] Although CTM as currently planned would not have this capability, the committee sees no technical reason why it could not, if this was considered a serious problem. (The objection to self-destruction mechanisms in nuclear missiles—that the target country might be able to exploit them—does not apply nearly so strongly to limited conventional strikes.)

The limited risk associated with debris can be illustrated with a simple calculation. For most systems, the debris resulting from normal operation or system

[2]Self-destruction would not eliminate debris; it would turn the large warhead into many small fragments. If the self-destruction occurred after the boost phase was completed, these fragments would continue along the same path as that which the intact warhead would have traveled. Although the total mass of debris that impacted the ground would be the same, the impacts would be spread over a much larger area and would occur at much lower velocities. This could reduce the potential for property damage and a high-casualty event such as the impact of an intact warhead on a school or hospital, but a complete assessment would require further study.

failure would have the potential to cause deaths out to a distance of perhaps 10 meters from the point of impact, corresponding to a lethal area of a few hundred square meters. The average population density of Earth is about 40 persons per square kilometer, so the expected number of deaths from a random impact is on the order of 0.01, or 1 death per 100 impacts. Given diversion capabilities that caused debris to land in less-populated areas, the expected damage would become much smaller. Thus, the risks associated with debris are unlikely to be a significant concern in situations serious enough to trigger a CPGS strike. Most uses of military force carry some risk to civilians, sometimes in third countries, as in the 1998 strike against al-Qaeda training camps in Afghanistan, which resulted in the crash of cruise missiles in Pakistan (though no casualties resulted from these failures).

ACCESS TO FORWARD BASING

Conventional Trident Modification and most other proposed CPGS concepts need not be deployed on forward U.S. military bases. This fact has led some proponents to suggest that CPGS systems could substantially compensate for or hedge against a loss of access to bases and other facilities in foreign countries, or even that they could provide a basis for curtailing forward deployments.

The problems for overall U.S. military capability that would arise from such reduced access are much greater than could be addressed by CPGS systems. In addition, some CPGS concepts depend to one degree or another on access to forward bases, as do many of the enablers (e.g., intelligence derived from unmanned aerial vehicles [UAVs] or from Special Operations Forces). Thus, the United States will retain strong interests in forward basing regardless of the deployment of CPGS systems.

Some CPGS options, such as CTM, are often said to be truly global in a way that forward-based options cannot be. But forward basing in regions such as the Middle East, where targets for CPGS are most likely to be located, will remain important not because such basing makes rapid attack possible, although that will sometimes be the case, but because the very interests that make CPGS potentially useful in those regions will also make it appropriate to maintain other military capabilities there. It would be a mistake to believe that CTM or other CPGS systems would make possible a meaningful reduction in forward basing.

Moreover, even if the United States goes forward with CTM, the most mature CPGS option, it will be several years until this capability is operational. Until that time, the United States will be entirely dependent on forward-based systems (fighter aircraft, armed UAVs, submarines armed with SLCMs, B-52s armed with ALCMs, tanker support for long-range aircraft based in the United States) to conduct strikes on distant targets. If the need for a CPGS capability is truly urgent, then it is no less urgent to maintain and, where possible, improve forward-based capabilities to carry out strikes against potential CPGS targets as rapidly as possible.

PROLIFERATION

Although there have long been dual-capable aircraft, the long-range ballistic missiles deployed by the nuclear-weapons states have been armed only with nuclear warheads. Attempts by other countries, such as North Korea and Iran, to acquire long-range ballistic missiles generally have been viewed as an integral component of an effort to acquire a nuclear-weapon capability. Such attempts have been opposed by the United States mainly on these grounds, sometimes with the argument that for reasons of cost and accuracy, ballistic missiles are inherently unsuited to conventional roles. The presumed connection between missiles and nuclear weapons is highlighted by the Missile Technology Control Regime (MTCR) guidelines, which begin as follows: "The purpose of these Guidelines is to limit the risks of proliferation of weapons of mass destruction (i.e., nuclear, chemical and biological weapons), by controlling transfers that could make a contribution to delivery systems (other than manned aircraft) for such weapons."[3] Although the text also refers to chemical and biological weapons, the 500-kilogram payload limit specified in the guidelines is based on the estimated mass of a nuclear warhead.

Some have argued that the deployment of CTM or other CPGS systems will break the presumed link between ballistic missiles and nuclear weapons, resulting in the spread of long-range missiles to additional countries. In this view, the deployment of a ballistic CPGS system by the United States would allow other countries to claim more credibly that their long-range missile programs were intended for conventional, not nuclear, delivery and are just as legitimate as the U.S. CPGS system. Similarly, it is argued that a weakening of the presumed coupling between missiles and nuclear weapons would weaken the MTCR and related efforts to limit the spread of missile technology. Although the Nuclear Non-Proliferation Treaty recognizes the right of only five countries to possess nuclear weapons, many countries argue that this division between "haves" and "have nots" should not be extended into other realms. In this view, the MTCR derives much of its legitimacy from its connection to nuclear nonproliferation—a connection that would be weakened by CPGS.

There is also a related argument that the U.S. acquisition of CTM or other CPGS capability would encourage or legitimate efforts by other countries to develop equivalent systems. There are, for example, reports of Russian officers who, after criticizing U.S. plans, note that such a system would be useful for Russia as well. The systems that others might develop would not necessarily mimic U.S. CPGS systems in being designed to deliver small payloads with high

[3]See <http://www.mtcr.info/english/guidetext.htm>. Last accessed on March 27, 2008.

accuracy. Indeed, they might be intended for very different sorts of targets, such as satellites.

Finally, it is argued that to the extent that U.S. CPGS systems threaten important targets in potential adversary countries, these systems might also stimulate other countries' efforts to develop and deploy capabilities that might deter such attacks, such as nuclear and other weapons of mass destruction and associated delivery systems. This is sometimes countered by arguments that CPGS, by demonstrating the U.S. ability to destroy adversary missile and WMD sites rapidly, would dissuade countries from even attempting to field such capabilities.

On balance, the committee judges that the deployment or use of CTM or other CPGS systems is by itself unlikely to have a substantial effect on the proliferation of ballistic missiles or nuclear weapons or on the development of similar systems by other countries. In general, countries will do what is in their own national interest and within their technological capability and financial capacity, regardless of what the United States does—or does not do—about CPGS. CPGS should not serve as a significant additional stimulant for the acquisition of systems that could attack high-value targets in the United States (e.g., in order to deter U.S. use of a CPGS system), because CPGS would not add meaningfully to the already very substantial U.S. strike capabilities that potential adversaries must take into account. Nor is CPGS likely to have a substantial effect on efforts to impede the spread of ballistic missile or nuclear technology. Countries that possess the technological capacity and have the strategic need for these capabilities have powerful incentives and self-interests in limiting the spread of these technologies. It is difficult to see how either developing CPGS or forbearing to do so could alter this to any significant degree.

ARMS CONTROL AND TREATY ISSUES

Two arms control agreements that are currently in force impose restrictions on the deployment of intermediate-range and long-range delivery systems: the Intermediate-Range Nuclear Forces (INF) Treaty of 1987 and the Strategic Arms Reduction Treaty of 1991 (START). The Moscow Treaty of 2002 (Strategic Offensive Reductions Treaty, or SORT) also requires reduction and limitation of strategic nuclear warheads. Below is a discussion of the impact of these treaty restrictions on CTM and other possible CPGS systems.

Intermediate-Range Nuclear Forces Treaty of 1987

The Intermediate-Range Nuclear Forces Treaty was signed by the United States and the Soviet Union on December 8, 1987. After the breakup of the Soviet Union, Russia, Belarus, Kazakhstan, and Ukraine became signatories and assumed the obligations of the former Soviet Union. The INF Treaty prohibits flight-testing, production, and deployment of all ground-based ballistic missiles and ground-launched cruise missiles with ranges between 500 and 5,500 kilometers, regardless of warhead type. The prohibition applies to missiles deployed anywhere in the world; thus, the United States could not deploy a conventionally armed intermediate-range ballistic missile or ground-launched cruise missile at a forward base, such as Diego Garcia in the Indian Ocean.

The INF Treaty is of unlimited duration. Accordingly, its restrictions remain in force unless the United States or Russia determines that extraordinary events have jeopardized its supreme national interests and exercises the right to withdraw from the treaty.

The INF Treaty places no restrictions on manned aircraft, air-launched or sea-launched systems, or ground-launched systems with ranges less than 500 kilometers or greater than 5,500 kilometers.

Strategic Arms Reduction Treaty of 1991

The Strategic Arms Reduction Treaty, which was signed on July 31, 1991, requires the United States and Russia to limit their deployed strategic arsenals to no more than 6,000 warheads, with no more than 4,900 on ICBMs and SLBMs, in accordance with agreed counting rules. The agreement limits deployed warheads by limiting the number of delivery vehicles and associated launchers to 1,600. After the breakup of the Soviet Union, a protocol to the treaty made Russia, Ukraine, Kazakhstan, and Belarus parties to the treaty and required the latter three to transfer their nuclear weapons to Russia. The required START reductions were completed by all parties in December 2001.

START covers all ICBMs and SLBMs and their associated launchers, including new types of ballistic missiles, and no distinction is made between nuclear-armed and non-nuclear-armed ballistic missiles. Thus, the submarine-launched CTM and SLGSM missiles and launchers would be subject to treaty limits. For example, each CTM would be counted as one SLBM launcher and eight warheads toward the aggregate limit on launchers and warheads. This should not present any practical problem, however, because under the Moscow Treaty of 2002 (see below), the United States plans to reduce the number of launchers and attributed warheads well below the START limits. The CTM missiles would, however, be subject to all START rules, including reentry vehicle on-site inspections to confirm that they are not armed with more than the permitted number of RVs. SLGSM would be handled differently from CTM. Presumably it would have to be either exempted on the basis of an agreed verification scheme or treated as a strategic weapon system (with an attributed number of RVs and an agreed definition of launchers, and so on).

Similarly, deployments of conventionally armed ICBMs would be covered by START. In addition to the limits on launchers and warheads noted above, START requires that ICBMs be deployed only in silo, road-mobile, or rail-mobile launchers; soft launchers are prohibited except at test ranges and space-launch facilities. ICBM bases must be located more than 100 kilometers from test ranges and space-launch facilities. Test ranges are limited to no more than 25 missiles and 20 fixed, soft launchers, which would restrict the number of conventional ICBMs that could be deployed at Vandenberg Air Force Base, California. Finally, START prohibits deployments of ICBMs outside national territory, as well as air-launched and surface-ship-launched ballistic missiles. These restrictions and prohibitions make it difficult for the United States to deploy conventional ICBMs while avoiding overflight and ambiguity problems that would arise from deployments at existing ICBM bases.

START defines a ballistic missile as a "vehicle that has a ballistic trajectory over most of its flight path." Therefore, boost-glide systems (even if derived from existing ICBMs or SLBMs) may not be subject to START if the glide portion of their flight is more than half of the total flight path. As noted in Appendix G of this report, however, the glide range generally may be much less than half of the total flight path, unless the initial speed is comparable with ICBM or orbital speed. However, at some small sacrifice in payload, glide range could be increased to exceed the ballistic range considerably by boosting the weapon to near-orbital speed.

Conventionally armed, long-range, air-launched cruise missiles are not restricted by START so long as they are distinguishable from nuclear-armed, long-range, air-launched cruise missiles.

Moscow Treaty of 2002

On May 24, 2002, the United States and Russia signed the Strategic Offensive Reductions Treaty (SORT) in Moscow. Under SORT, they agreed to reduce the number of operationally deployed strategic offensive warheads to 2,200 or fewer by December 31, 2012. The warhead limit takes effect and expires on the same day. The treaty contains no verification provisions or agreed counting rules. Because SORT applies only to nuclear warheads, it imposes no restrictions on ballistic missiles or other delivery vehicles armed with conventional warheads. Thus, SORT should have no effect on CTM or any other CPGS system.

Future Agreements

START is scheduled to remain in force until December 5, 2009. The parties are committed to meet at least 1 year before this date to consider whether the treaty should be extended. Although START contains provisions for an indefinite series of 5-year extensions, neither the United States nor Russia has indicated that it wishes to extend START in its current form. This has led some observers to conclude that the United States should simply ignore the restrictions that START would impose on CPGS systems, because START will expire before CPGS systems could be deployed.

The START restrictions are unlikely to disappear entirely, however; the United States and Russia are likely to attempt to extend some of its provisions or to replace it with another agreement. One issue in any such negotiation would be the degree to which the United States would insist on modifications or provisions designed to protect certain CPGS systems or deployment options. As in any negotiation, this would have to be weighed against other goals for both parties.

A key issue in the negotiation of a follow-on agreement would be verification and transparency measures. Because the central goal of such an agreement would be to limit nuclear weapons, there might be a temptation to try to exclude conventionally armed ballistic missiles from any such measures. Such an exclusion would lead to another kind of ambiguity problem and seriously complicate any effort to verify limits on nuclear systems. It will be desirable to reassure Russia (and others) that CPGS systems are not armed with nuclear weapons by using transparency measures that could be more intrusive and revealing than would be acceptable with nuclear systems.

Looking to the longer-term future, the existence of substantial numbers of CPGS systems might be a concern because of the possibility that they could be rearmed with nuclear weapons. This would not be a concern at the relatively high limits on nuclear warheads established by START or even the Moscow Treaty, but it could arise if the United States and Russia move to significantly lower levels—for example, less than four times the number of long-range conventional-missile warheads. Such concerns could, however, likely be resolved with transparency measures to reassure other parties that conventional missiles are not armed with nuclear warheads (and, perhaps, that additional nuclear warheads are not available to arm the conventional missiles).

Thus, assuming that the United States is willing to accept far-reaching transparency measures—which it ought to be, for reasons related to resolving ambiguity concerns—CPGS should neither be a barrier to future arms control agreements nor a reason not to pursue future arms control agreements.

STRATEGIC CONSIDERATIONS

A significant fraction of Russian strategic nuclear forces—submarines in port and long-range bombers in soft shelters—could be vulnerable to precision attacks with conventional weapons. If penetrating warheads capable of destroying hardened surface targets are available, some command, control, and communications facilities and, with very high accuracy, missile silos, could be vulnerable. Even if the initial deployment is very limited, any CPGS system could in principle be substantially expanded after its development and testing have been successfully completed. Such expansion would raise at least the theoretical prospect of a conventional preemptive strike capability against Russian nuclear forces. These concerns would be more acute for China, because China's nuclear forces are much smaller and less diverse than those of Russia. However, even for China, large numbers—at least many hundreds—of CPGS weapons would be needed to mount an even marginally plausible conventional threat to a sophisticated nuclear force. Support for and deployment of any of the CPGS systems considered in this report would not imply that such a massive "strategic" CPGS system should be built, and the committee would not endorse such a project.

If the United States were perceived by Russia or China to have embarked on such a course, they could respond to the development of this capability in many ways. In an attempt to strengthen deterrence against such attacks, they might, for example, declare that they would respond to a conventional attack on their nuclear forces and other strategic assets as if it were a nuclear attack. For China, this would represent a major change from its current declaratory policy of no-first-use of nuclear weapons, but it would be a change that China might well make as a response to a clear threat to its deterrent.

Russian or Chinese leaders might believe, however, that threats to order nuclear retaliation to conventional attacks would be less than credible in the eyes of U.S. leaders and that CPGS might therefore reduce their ability to deter U.S. attacks. It would be reasonable to expect that Russia and China would respond to the prospect of a large-scale CPGS system by moving, as rapidly as technology and resources would permit, to decrease the vulnerability of their nuclear forces to precision conventional attacks. Such efforts might include the following actions:

- Increasing the number or readiness of mobile launchers—mobile ICBMs and submarines at sea—which are difficult or impossible to target with CPGS systems;
- Expanding the total number of deployed launchers or warheads (or refusing to implement already-agreed reductions) in order to ensure the survival of a given number of warheads after a conventional attack;
- Hardening facilities to make them less vulnerable to conventional attack;
- Developing close-in defensive systems to protect point targets, such as ICBM silos, from conventional attack;
- Enhancing the ability to launch vulnerable nuclear forces rapidly on warning of, or in the midst of, a conventional attack; and
- Using cover and deception to disguise or misrepresent the locations of their systems.

Some of these actions, such as increasing readiness or the ability to launch on warning, if taken by Russia or China, could increase the probability of mistaken attacks on the United States.

It should be emphasized that there would be serious practical obstacles to expanding a CPGS system to the point where it might be viewed by Russia as having a significant capability against its nuclear forces, even if the capacity to destroy hard targets had already been developed and demonstrated. The largest barrier is deploying the many hundreds or even thousands of missiles that would be required to mount a significant conventional threat against Russian nuclear forces. This could be accomplished in less than a decade only if many, if not most, of the existing nuclear-armed Minuteman III and Trident II missiles were converted to the CPGS mission. But such conversions would simultaneously draw down U.S. nuclear forces, with the result that the overall threat perceived by Russia would not increase and might even decrease. Moreover, Russia is unlikely to consider plausible a scenario in which the United States would sacrifice a large fraction of its nuclear deterrent force to the CPGS mission.

Thus, as a practical matter, deploying a CPGS system that could threaten Russian nuclear forces would likely involve building at least several hundred additional intercontinental-range missiles and their associated launchers.[4] The additional cost of the missiles alone would be several tens of billions of dollars; launchers would add billions more. This would be a highly visible project that would take more than a decade to carry out, giving Russia plenty of time to consider its response. Thus, even conservative Russian planners would have no reason to react to the limited CPGS deployments considered here by modifying their nuclear forces or associated policies. They could instead monitor the situation, while making clear to the United States that larger deployments would be a matter of serious concern.

As discussed elsewhere, the United States can take steps intended to assure Russia that it is not the intended target of a CPGS system and that the system has little capacity to destroy certain types of targets. In addition to keeping the number of deployed CPGS systems small, the United States can limit CPGS deployments to areas that minimize the potential for attacks deep in Russian territory, and it can adopt transparency measures, such as launch notification, joint early warning, technical briefings, inspections, and devices integrated into the payload that signal a conventional missile launch.

At the same time, U.S. decision makers should be aware of the possible reactions of Russia and China—and the potential impact of these reactions on U.S. security—to a decision to move from a limited CPGS capability, such as that of CTM, to one capable of high-volume attacks that could possibly threaten Russian or Chinese nuclear forces. For this, and many other reasons, the United States should, in deciding to proceed with any CPGS program, make clear both in its declaratory policy and in its procurement actions and doctrinal development that it is not embarking on any such far-reaching effort.

[4]A possible exception would be the redeployment in a CPGS mode of some or all of the Minuteman II missiles that were withdrawn from deployment and their silos launchers that were destroyed under START. These missiles are currently in storage in the Rocket Systems Launch Program and could be redeployed, but this would require building new silo launchers, which would require many years and billions of dollars to complete. In addition, it would be necessary to identify and clarify any START restrictions on redeploying a missile (such as Minuteman II) which has been designated under START as a "retired missile."

4

Technology Issues

OVERVIEW

The key goal for conventional prompt global strike (CPGS) systems, the ability to hit distant targets quickly and accurately, delivering damage only to the well-defined target area, entails serious technical challenges. A wide range of concepts has been proposed to meet the CPGS challenge, as illustrated in Figure 4-1. These solutions include modest modifications to existing ballistic missiles (such as the Conventional Trident Modification [CTM]), boost-glide missiles, and hypersonic cruise missiles. The weapons themselves can be launched from the continental United States (CONUS) or from forward-deployed land bases, submarines, or aircraft.

The principal proposals for CPGS build on technology designed for the rapid, long-range ballistic delivery of nuclear weapons. Since a conventionally armed prompt global strike system must be much more precise in its targeting accuracy and weapon delivery than a nuclear delivery system, no existing system can be used without modification. CPGS systems are being considered with varying degrees of modification from a ballistic trajectory, as illustrated in Figure 4-2. The simplest modification requires developing sufficient control, navigation, and guidance during reentry to enable the required terminal accuracy, as in the left-most options shown in Figure 4-2. For some options a gliding reentry vehicle (RV) is used to increase range, trajectory flexibility, and maneuverability, as illustrated in the two right-most options shown in Figure 4-2. In addition to the terminal accuracy challenge, these approaches also must solve the serious issues of managing the heat generated when traveling at high speed within the atmosphere.

A wide variety of CPGS payloads has also been proposed for use, as illus-

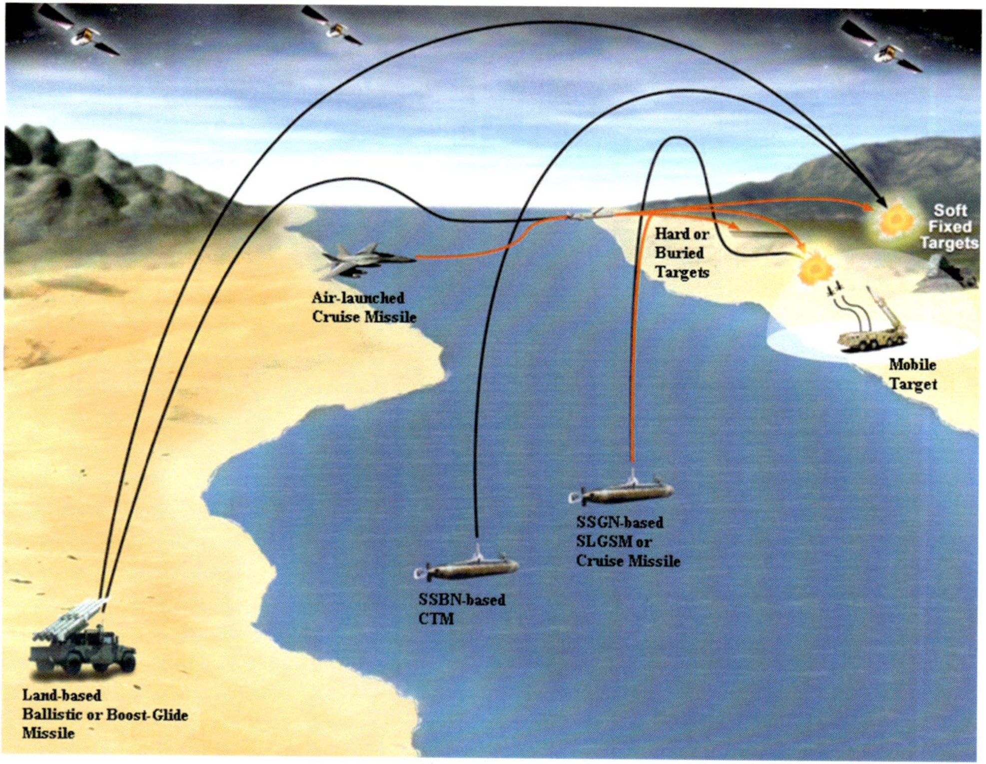

FIGURE 4-1 Candidate concepts for conventional prompt global strike include ballistic missiles, boost-glide vehicles, and hypersonic cruise missiles. The range of these concepts varies from 1,000 to 3,000 nmi for forward-deployed systems to more than 10,000 nmi for land-based systems. Options are being investigated to attack a range of targets from soft to hard to mobile. NOTE: Acronyms are defined in Appendix A.

FIGURE 4-2 Illustration of the reentry vehicles (RVs) proposed for different stages of conventional prompt global strike (CPGS) systems. A previously developed modification to ballistic reentry, E2, is the basis for the proposed short-term Conventional Trident Modification (CTM) option. For the Submarine-Launched Global Strike Missile (SLGSM), a scaled-up version of the previously developed Mk 500 is the proposed RV, which is designed to have a glide range less than 1,000 nmi. The Conventional Strike Missile (CSM)-1 concept builds on theAMaRV vehicle and is considered to have an 800-second glide segment to its trajectory. The CSM-2 concept builds on a high lift-to-drag (L/D) vehicle, which is being developed by the Defense Advanced Research Projects Agency FALCON program, that has a 3,000-second thermal protection system (TPS). NOTE: AMaRV, advanced maneuvering reentry vehicle; E2, Enhanced Effectiveness (one of two test beds for demonstrating proof-of-principle concepts for ballistic missile delivery in CPGS and discussed in Chapter 4 in the subsection entitled "Guidance, Navigation, and Control Accuracy Issues"); LETB, Life Extension Test Bed (the second of two test beds for demonstrating proof-of-principle concepts for ballistic missile delivery in CPGS and discussed in Chapter 4 in the subsection entitled on "Guidance, Navigation, and Control Accuracy Issues").

trated in Figure 4-3. In the terminal phase of flight, the RV or cruise missile may attack the target with a unitary warhead (which may be designed for penetration), or a warhead that disperses kinetic energy projectiles (KEPs). The reentry vehicle may fly a nominally ballistic flight path with entry angles of less than 30°, or the reentry vehicle may maneuver aerodynamically for vertical attack of the target. Alternatively, the delivery platform may slow sufficiently to deploy existing munitions, sensors, communication relays, or unmanned aerial vehicles (UAVs). The dispensing sequence may occur at low speed following deceleration of a low-β reentry or may be required to occur at supersonic or hypersonic speeds for survivability reasons.

The following sections review the CPGS options and technology challenges, including the system requirements, system concepts, research and development (R&D) issues, and technology readiness time lines.

REQUIREMENTS

In its most general definition, the mission of a CPGS system is to provide the capability to attack and defeat with conventional weapons a time-sensitive target anywhere in the world. In considering a CPGS capability, one must address the complete end-to-end system-of-systems capability to identify potential targets, determine their geolocation, define attack options, estimate collateral damage, communicate effectively with a decision maker who chooses to go forward with the strike, deconflict flight through air and space domains, engage the target, and assess the impact of the attack. The time required to complete the entire target prosecution time line must lie within the target's period of vulnerability. A version of the attack time line is shown in Figure 4-4 as consisting of the following phases: Find, Fix, Track, Target, Engage, and Assess (F2T2EA). As illustrated in Figure 4-4, the development of a robust and flexible CPGS capability requires shortening all aspects of the target prosecution time line. With respect to the engagement portion of the F2T2EA process, CPGS options that have the potential to engage a target within 1 hour are being explored.

Compression of the time line to the extent required will necessitate seamless integration of disparate systems; development of detailed tactics, techniques, and procedures (TTPs); and personnel training. It appears likely that time lines consistent with the CPGS mission can be achieved, but additional development work will be required. For the near to mid-term, some of the steps will need to have been accomplished in advance of when the 1-hour clock begins. In the mid-to-long term, improved procedures and technology may make it possible that the entire process could be accomplished within the hour under some conditions. In fact, in examining several scenarios and looking at historical cases analogous to what is envisioned (see Chapter 2), the committee found that many of the missions postulated for this capability can largely be preplanned on a contingency basis, reducing the stress on these enabling activities. What cannot be preplanned is the

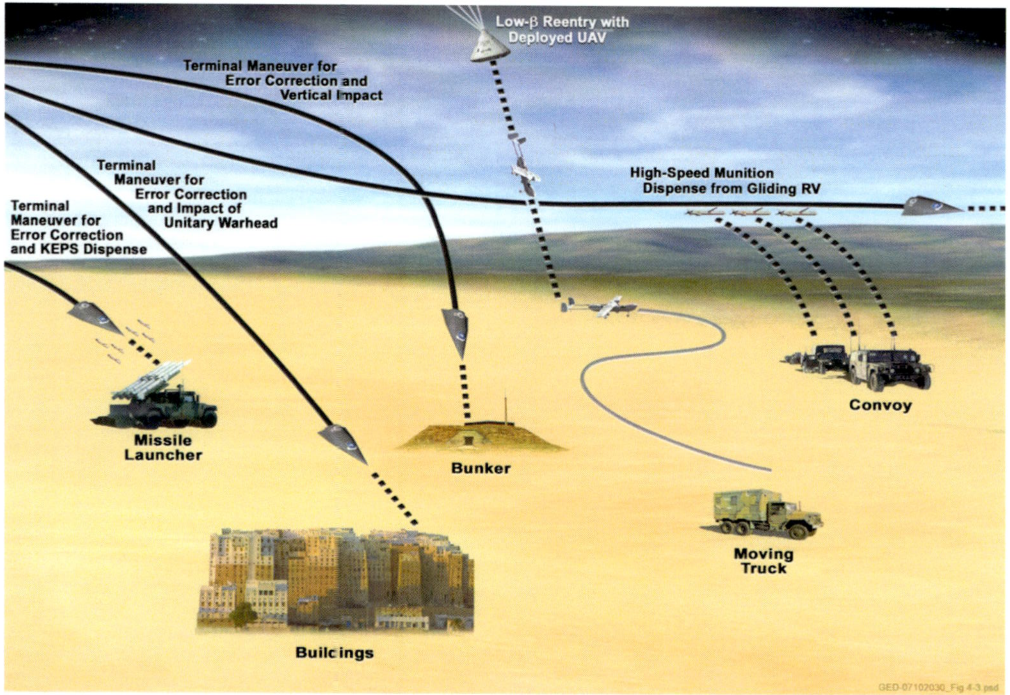

FIGURE 4-3 The need to attack different types of targets drives a requirement for different terminal-phase maneuvering and warhead characteristics. Terminal-phase options for conventional prompt global strike include minimal error corrections to the ballistic trajectory to achieve terminal accuracy, aerodynamic maneuvering for vertical impact to allow attack of buried structures, long-range glide for range extension and potential dispensing of munitions, and low-β reentry with deployment of an unmanned aerial vehicle (UAV) for attack of a moving target. NOTE: Acronyms are defined in Appendix A.

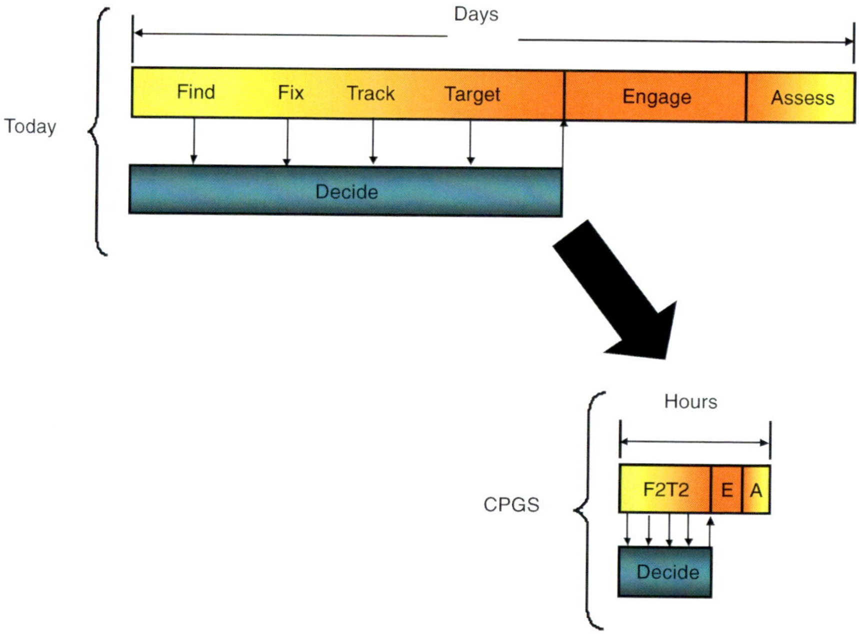

FIGURE 4-4 Conventional prompt global strike (CPGS) will require compression of the time line across the entire Find, Fix, Track, Target, Engage, and Assess (F2T2EA) strike process. With strategic warning and preparation, key aspects of this process can be accomplished in advance. For those portions of the process that need to occur inside the compressed time, technical means for seamlessly integrating systems that are currently not interoperable will need to be developed.

decision and validation time when the contingency actually occurs. The speed of execution time required in CPGS decision and validation, including the time for earlier steps in those limited cases when preplanning has not occurred, exacerbates an already difficult command, control, communications, computers, intelligence, surveillance, and reconnaissance (C4ISR) situation for the U.S. military, as discussed in a recent report of the National Research Council.[1]

The "global" requirement for a CPGS system places significant requirements on potential engagement systems, one that the committee concluded should not be taken literally.[2] The "prompt" requirement for a CPGS system also significantly

[1]National Research Council. 2006. *C4ISR for Future Naval Strike Groups.* The National Academies Press, Washington, D.C.

[2]If "global coverage" were free, it would not be an issue, but the committee found, from its own analysis, that modest restrictions of coverage (e.g., not worrying about a sudden need to fire on Antarctica) would in some cases allow material improvements in payload and more flexibility in system choice.

impacts the design characteristics of potential engagement systems. In assessing the requirements for CPGS, it is important that the terms "global" and "prompt" be viewed in the context of the realistic time lines associated with gathering and evaluating credible intelligence, surveillance, and reconnaissance (ISR) data, planning of force packages for a precision strike, and the decision process authorizing the strike. The range of scenarios considered in this report is intended to illuminate the full spectrum of mission options.

The concept of a prompt global strike capability initially evokes ideas involving strategic ballistic missiles—either land- or sea-based—that can deliver payloads over enemy defenses to targets 6,000 or more nautical miles away within 30 to 40 minutes of their launch. If one defines the term "global strike" to mean the ability to strike anywhere in the world without depending on basing or overflight rights, but perhaps with foresight in positioning launch platforms, one is led to look at approaches other than intercontinental ballistic delivery. In this approach, "prompt" might be measured from the time of decision or target-cue to the time of weapon effect. That is, preparations and decisions would have been made in advance.

A number of technical system options are potentially available for CPGS. Specifically, platforms (e.g., submarines, land-based aircraft, carrier-based aircraft) can be moved forward to an area of interest and can loiter (in some cases covertly) with long persistence—for many days at a time—while the planning and approvals for the force package are proceeding in parallel. These platforms can carry fast-flying shorter-range stand-off weapons that have flight times of 30 minutes or less to arrival on target. Armed remotely piloted vehicles have been employed on several occasions and represent one end of the spectrum of this concept, as does a cruise missile with loiter capability, but both have far less loiter endurance than needed and are likely to be detected while they loiter. A naval battle group (with aircraft, cruise missiles, or both) can sometimes run undetected in an area for days or weeks, but it might be observed at any time. In contrast, a nuclear-powered attack submarine (SSN) or a nuclear-powered ballistic missile submarine (SSBN) deployed in a region of interest with medium-range ballistic missiles would meet the criteria of covertness and persistence without significant limitations, if necessary technical capabilities are developed and weapon effectiveness is demonstrated.

Two other important desirable characteristics are embodied in the CPGS concept. One is the need to have a high probability of penetrating defenses and reaching the target, and the other is the ability to do this without risking U.S. personnel. These concerns are highest in the leading edge of an attack against a formidable adversary when the adversary's air defenses have not been suppressed.

These considerations favor longer-range ballistic missiles or boost-glide missiles or hypersonic cruise missiles that could be launched from submarines, long-range aircraft, or land bases. Some systems that would launch from CONUS, however, might be constrained by concerns about overflight (e.g., a missile overflying Europe, Russia, or China) and about debris from booster stages (which might, for example, fall into Canada). Whatever scheme is proposed, there are ambiguity issues, potential compromise of other mission capabilities, and international policy and arms stability questions to be considered (see Chapter 3).

In discussing the range requirement for a CPGS system, it is helpful to look at the geography of the world overlaid with range contours and azimuths to and from various locations. The (nonrotating Earth) ranges[3] from the coasts of the United States (where CONUS-based systems may be based) to various parts of the world are shown in Figure 4-5. One can see that a CONUS-based CPGS system must have a range of 6,500 to 7,000 nmi to reach most parts of the world if ballistic trajectories are flown. If overflight considerations constrain operations, the required range may be as much as 16,000 nmi, since the system may be required to fly the long way around.

Figure 4-6 shows ranges from a potential operating region in the Arabian Sea. One can see that a CPGS system operating from this area requires a range of approximately 2,500 to 3,500 nmi to reach most of the troubled regions of the area.

With range requirements defined, the speed requirements for candidate concepts are shown in Figure 4-7, where the capabilities of existing conventional forces, hypersonic cruise missiles, and ballistic missiles are overlaid on lines of average Mach number (defined as the range divided by the time-to-target and stratospheric sound speed, assumed to be 968 ft/sec).[4] The existing capability for conventional force projection is limited to subsonic speeds. With a 1-hour engagement window, this limits the range to approximately 460 nmi.

The range-time trade-off for ballistic missiles is also shown in Figure 4-7. In this case, the time-to-target was calculated assuming maximum-range ballistic flight for each range. Ballistic missiles can engage targets within the 1-hour engagement window for both forward-deployed and CONUS-based systems. From forward-deployed platforms, ballistic missiles can engage targets in approximately 25 minutes, whereas CONUS-based ballistic missiles require a flight time of approximately 40 minutes.

Finally, the capabilities of hypersonic cruise missiles are shown corresponding to flight at speeds of between Mach 4 and Mach 8. (In this report, the term "hypersonic cruise missiles" will be used to refer to missiles that cruise at speeds at or above Mach 4.) Even hypersonic air-breathing cruise missiles are not capable of deployment from CONUS in the 1-hour engagement window, but they can be deployed from forward-deployed forces with flight times between 30 and 60 minutes.

---

[3]Unless specifically expressed otherwise, the range contours in this report are presented as nonrotating Earth ranges.

[4]Mach 1 defined in this context as the speed of sound in the stratosphere = 968 ft/sec (or equivalently 0.159 nmi/sec, or, 1,062 km/hr).

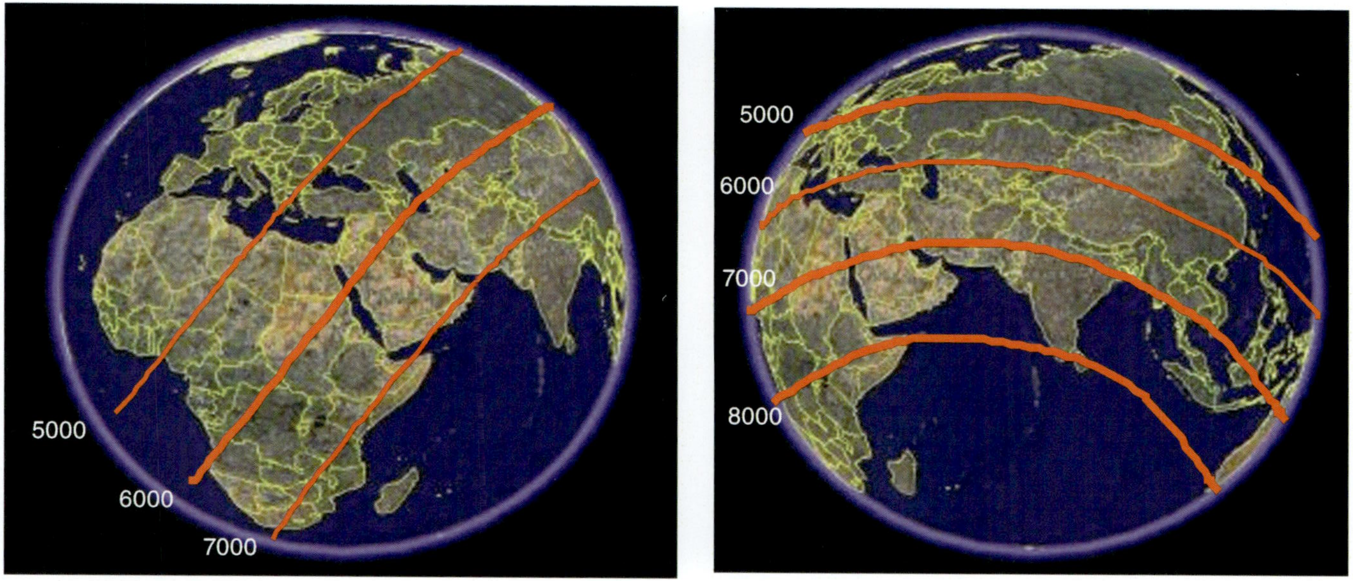

FIGURE 4-5 Sample great circle ranges (nmi) from (left) Kennedy Space Center, Florida, and (right) Vandenberg Air Force Base, California, illustrating the need for 6,500 to 7,000 nmi to reach most parts of the world for a continental United States–based system. Overflight restrictions can result in significantly longer-range requirements, since the vehicle may be required to travel the long way around the globe.

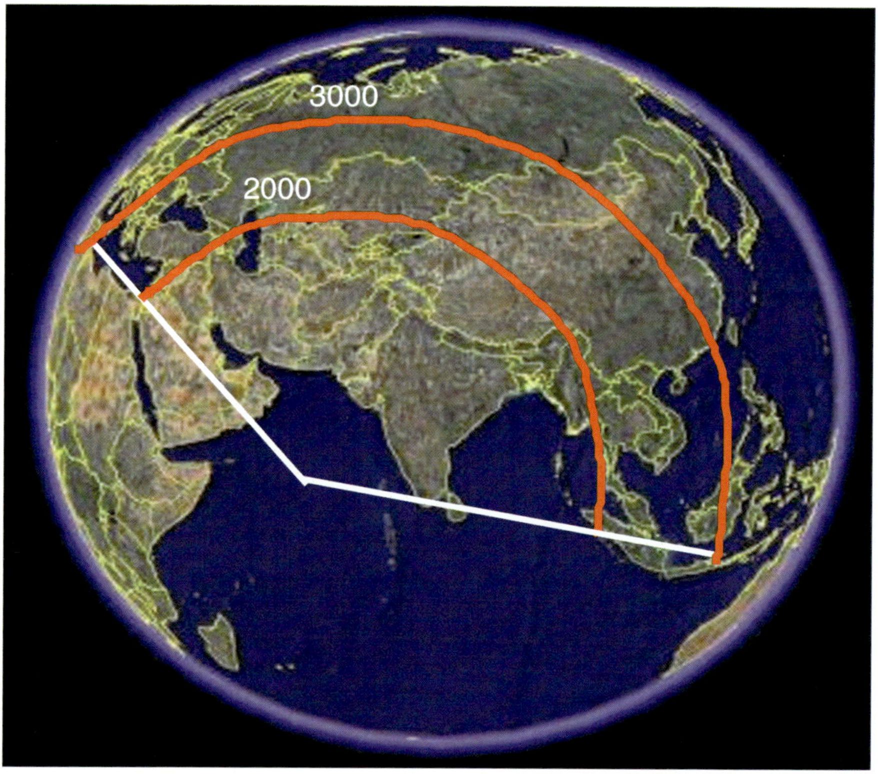

FIGURE 4-6 Sample ranges (in nautical miles) from a notional nuclear-powered guided missile submarine (SSGN) launch location in the Arabian Sea illustrating the need for a conventional prompt global strike range capability of 2,500 to 3,500 nmi when launched from an SSGN.

Target Types and Information Required

The targets sets that are most relevant to CPGS can be divided into three general categories: (1) fixed soft targets (both point and area targets), (2) fixed hard targets (both hardened point targets and deeply buried complexes), and (3) mobile targets or targets with uncertain location. Only the subset of targets within these general categories that are considered to be time-sensitive will be candidates for attack with a CPGS system.

Fixed Soft Targets

A fixed soft target is one whose geolocation is known with accuracy acceptable for attack, whose location remains constant during the engagement window,

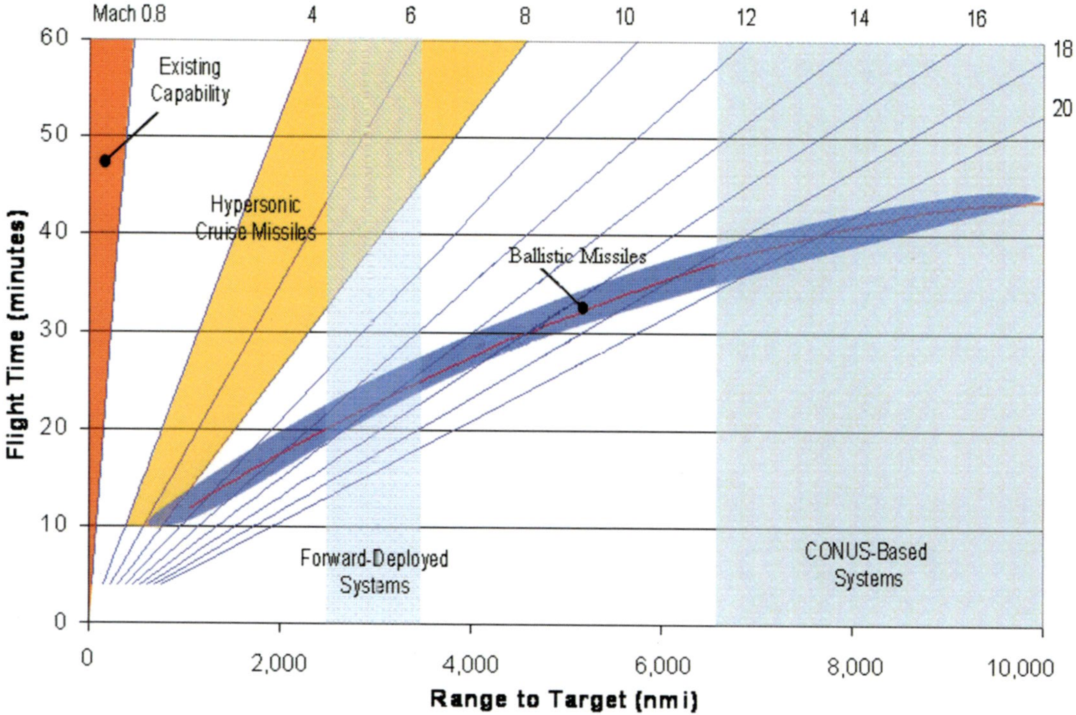

FIGURE 4-7 Flight time versus range for existing subsonic systems, proposed hypersonic cruise missiles, and ballistic missiles. Overlaid are lines of average Mach number, defined as the ratio of block speed (range divided by flight time) to the sound speed at 390 degrees Rankin. For continental United States (CONUS)-based systems, only ballistic missiles are capable of meeting a 1-hour time line. For forward-deployed systems, hypersonic cruise missiles are candidates.

and whose construction is sufficiently light that penetrating warheads are not required. Examples include personnel at a fixed location (potentially located in an non-reinforced building or structure), missiles loaded on gantries, and relocatable systems such as trucks, ships, air defenses, and missile launchers during time periods when they are in fixed positions. A special case of soft targets is the soft large-area case in which a personnel target is known to be in a given area that contains multiple buildings, tents, or vehicles, such as a training camp. If the target's location within the large area is not determined, then the probability of an effective hit would be reduced, subject to the area of uncertainty and the radius of dispersion of the KEPs. The use of multiple offset weapons would increase the odds of an effective strike, but at the cost of increased collateral damage.

For many parts of the world, those where the National Geospatial-Intelligence Agency (NGA) has built a Digital Point Positional Data Base (DPPDB), mission planners can geolocate targets within minutes with an accuracy measured in meters once the target has been found. However, if a target lies within an area not covered by a DPPDB, it can take a prolonged period to achieve the accurate geolocation. Greater efforts are needed to speed up this process. Specifically, NGA needs to accelerate the production of the DPPDB to extend its geographic coverage and develop a means to determine rapidly and accurately the geographic coordinates of any visually identified point (e.g., from a recent photo taken in the field) when that point lies outside DPPDB coverage. Whether or not the target lies in an area covered by a DPPDB, a principal additional challenge will often lie in intelligence efforts to find the target and verify the information indicating that a strike is warranted. In summary, the committee has evaluated technical enablers associated with finding and locating fixed, time-sensitive soft targets, as discussed in Chapter 2, and has concluded that some credible scenarios exist where a target could be found and geolocated in a time frame consistent with the CPGS mission.

Fixed Hard Point Targets

A fixed hard target is one whose geolocation is known with accuracy acceptable for attack and whose hardness or construction is sufficiently strong that a penetrating warhead is required. Examples include buried structures or hardened command-and-control (C2) bunkers and hardened aircraft and missile shelters. The payloads proposed for the earliest versions of a CPGS system have limited capability for seriously damaging such targets, as discussed in the subsection "Weapons Effectiveness," below. More capability will be available in later versions.

Some fixed hardened targets can be identified and geolocated prior to the period when they take on a time-sensitive urgency and some (those that lie in an area covered by a DPPDB) can be quickly geolocated once identified.

In summary, the committee evaluated technical enablers for this class of target and determined that, with sufficient planning, the time line associated with the upfront technical portions of the F2T2EA process could in some cases support the CPGS mission.

Mobile Targets or Targets with Uncertain Location

Both mobile targets and targets whose geolocation accuracy is insufficient to attack with conventional coordinate-seeking weapons present significant challenges to a CPGS system. One type of example includes personnel not known to be in a fixed location such as a particular meeting room (see above for soft large-area targets) or land vehicles and ships that are underway. A second type of example would be mobile intercontinental ballistic missile (ICBM) launchers that move to the field but then sit at poorly known fixed positions when firing weapons.

This class of target places significant extra technical requirements on a CPGS system, with different constraints for the two types of example. Three general approaches to overcoming this challenge have been considered in this study (others are possible): (1) continued observation combined with in-flight target updates, a topic not addressed in detail in this study; (2) the deployment of submunitions capable of independent target acquisition and prosecution; or (3) ballistically delivered UAVs with onboard sensors, communications, and weapons and sufficient endurance to search for, acquire, prosecute, and assess damage to targets. The committee believes that development of the capability to attack this class of target is important and that research, development, testing, and evaluation (RDT&E) in this area must continue.

Countermeasures

All approaches to targeting with CPGS will, of course, be subject to countermeasures, and some of the proposed CPGS options specifically address approaches to deal with countermeasures, such as the use of terrain to protect against attack. Continuing effort against countermeasures will be a military necessity as opponents become aware of CPGS capabilities and adopt tactics such as using terrain well, deception in locating things underground, using large underground facilities, "hiding" among civilians, jamming, decoys, and so on.

SYSTEM CONCEPTS

Strike missiles are held by developed military forces the world over. Figure 4-8 plots the range and gross launch weight of existing missiles, together with shaded regions indicating sizing trends related to CPGS. The shaded regions for subsonic and hypersonic cruise missiles and boost-glide vehicles are based on analyses conducted by the committee, which are described later in this chapter.

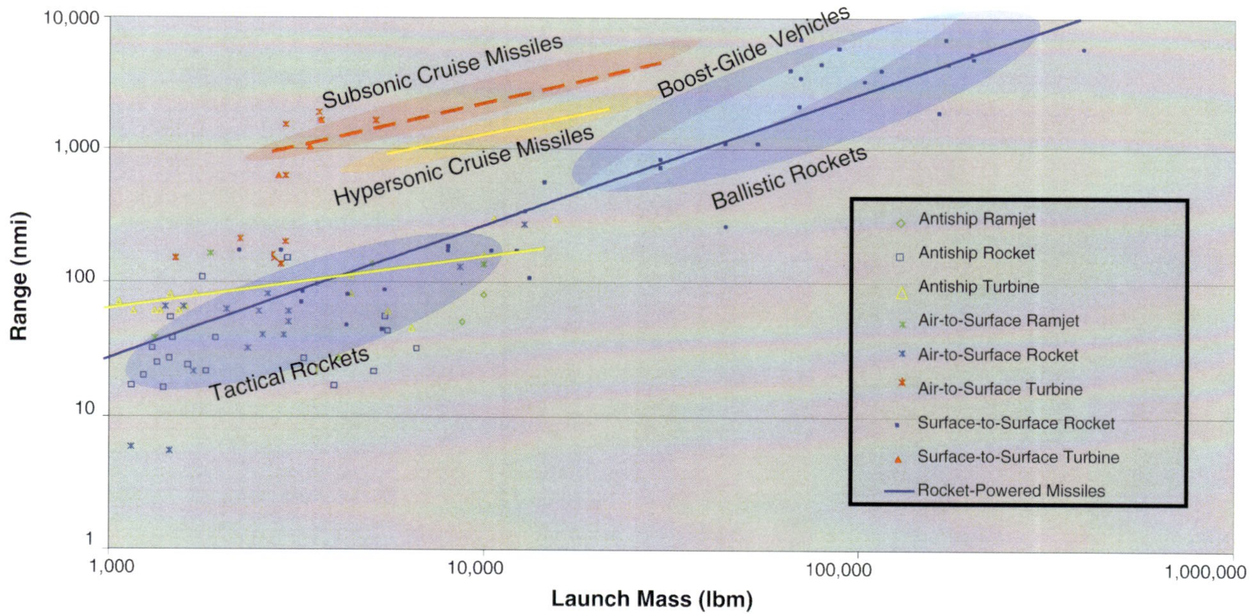

FIGURE 4-8 Worldwide inventory of missiles. The blue data points represent rocket-powered systems, which can be broadly divided into tactical and long-range ballistic missile domains. The light-blue shaded region designated "Boost-Glide Vehicles" illustrates the range benefit of gliding reentry vehicles compared with ballistic missiles. The red data points represent subsonic cruise missiles powered by turbojet engines. The committee considered the potential of larger subsonic cruise missiles, and the red shaded region represents the trend of range versus launch mass for subsonic cruise missiles. The committee also considered hypersonic cruise missiles. While no strategic hypersonic cruise missiles exist in the world currently, the trend of missile range versus launch mass is also shown. SOURCE: Based on data in *Aviation Week and Space Technology Aerospace Source Book 2006*, January 16, pp. 184-199.

While this presentation of missile capability hides important factors such as warhead weight, number of stages, and launch platform, several interesting trends can be observed. Missiles capable of operation from CONUS (range > 6,500 nmi) are ballistic or boost-glide in nature. For intermediate ranges (2,500 nmi < range < 3,500 nmi), ballistic, boost-glide, and hypersonic cruise missiles are all candidates.

A number of engagement system concepts to support the CPGS mission were presented to the committee by both government and industry sources. These systems can be broadly classified as ballistic missiles, boost-glide missiles, and hypersonic cruise missiles. The committee also considered some additional ideas. Of these, one system concept, which the committee refers to as CTM-2, seems quite worthy of more serious consideration, for reasons indicated in Chapter 2. The sections that follow provide brief descriptions of various system concepts.

Proposed Systems

The USAF Space Command is currently conducting an analysis of alternatives (AoA) for CPGS systems. Candidate systems under consideration within the AoA are outlined in Table 4-1. The Conventional Trident Modification is the only system (with the possible exception of CTM-2) capable of near-term deployment. The mid-term option, called the Conventional Strike Missile (CSM), serves as the basis for two alternatives discussed in this report, CSM-1 and CSM-2. Within the AoA, four long-term options are being considered: CONUS-based missile (similar to the committee-proposed CSM-2), forward-based missile, Submarine-Launched Global Strike Missile (SLGSM); and Mach 6 missile. These options are discussed in the subsections below.

As discussed below, CTM could also serve as the first step in the evolutionary development of technology for other high-speed delivery systems that offer additional capability and flexibility. In this development scenario, technology development would begin with a demonstration of the necessary control on limited-maneuverability ballistic reentry vehicles, followed by the development of maneuverable endoatmospheric reentry flight paths of increasing range that would support more flexible and powerful military options. This evolutionary path was suggested in Figure 4-2. The path draws from test trajectories: the Enhanced Effectiveness (E2) test, which is the basis for the proposed CTM; and the Life Extension Test Bed (LETB), which is the basis for the proposed SLGSM and for proposed developments involving longer reentry flight paths, as discussed below in the subsections on SLGSM and CSM-1 systems (see below the subsection on "Guidance, Navigation, and Control Accuracy Issues" for a discussion of the E2 and LETB tests). The final reentry vehicle type shown (designated CSM-2) minimizes the initial ballistic phase and exploits long-range hypersonic glide. This option is discussed in the subsections below on CONUS-based boost-glide missiles and advanced hypersonic weapons.

TABLE 4-1 Potential Candidates for Achieving Conventional Prompt Global Strike Capability Used in the Air Force Space Command Analysis of Alternatives (AoA)

	Near Term	Mid-Term	Materiel Solutions Examined in PGS AoA (Long Term)			
	Conventional Trident Modification	Conventional Strike Missile	CONUS Missile	Forward-Based Missile	SLGSM	Mach 6 Missile
Projected IOC	FY10 Funded FY08 President's Budget	FY14/15 Demo Program Funded FY08 President's Budget	FY20 Conceptual (No funding)	FY20 Conceptual (No funding)	FY20 Conceptual (No funding)	FY20 Conceptual (No funding)
Launch Vehicle	Trident D-5	Minotaur III Class Booster	Minotaur III Class Booster[a]	Minuteman III Class Booster	Two-stage Solid Rocket Booster	B-52
Payload Delivery Vehicle	Modified Mk-4 (68 in.)	Biconic Hypersonic Glide Vehicle (161 in.)	Biconic Hypersonic Glide Vehicle (161 in.)	Biconic Hypersonic Glide Vehicle (161 in.)	Biconic Medium Re-entry Vehicle (96 in.)	Cruise Missile (Scram Jet with Rocket Booster)

Max Range	Classified	10,300 nm [b]	9,000 nmi+	4,200 nmi	3,500 nmi with 1,150 lb Payload	2,000 to 2,800 nmi[c]
Payload Capacity	1,000 lb (4 250 Mk-4s)	2,000 lb	2,000 lb	2,000 lb	900 lb	1,000 to 2,000 lb
Payload Volume	1.25 ft^3	22 ft^3	22 ft^3	22 ft^3	6 ft^3	22 ft^3
Potential Payload Concepts	KEP	Dispensed Munitions, KEP, Penetrator	Dispensed Munitions, KEP, Unitary Blast Fragments or Penetrator	Dispensed Munitions, KEP, Unitary Blast Fragments or Penetrator	KEP, Unitary Blast Fragments or Penetrator	Dispensed Munitions, KEP, Unitary Blast Fragments or Penetrator

NOTE: The mid-term Conventional Strike Missile and long-term CONUS missile concepts are similar to the CSM-1 and CSM-2 in this report. Acronyms are defined in Appendix A. This table, taken from the Air Force's analysis of alternatives, is shown for comparative information purposes only. The committee's configurations and assessments of alternate CPGS systems differ in some cases.

[a]Different booster stack from CSM.

[b]18,600 nmi with advanced TPS.

[c]Does not include B-52 fly-out range extension.

SOURCE: Maj. Steven Kravitsky, USAF, Headquarters, Air Force Space Command, "Prompt Global Strike Payload Development," presentation to the committee, May 11, 2007, Washington, D.C.

Sea-Based Missiles

Figure 4-9 shows the time line for near-term and longer-term CPGS system options that could be deployed on a submarine platform. The baseline system is the CTM, which could be fielded by 2011 depending on funding decisions in 2008. CTM-2 (not shown in Figure 4-9) could be a second step in CTM's evolutionary path, or a first-step development itself. The development and demonstration of technology to meet the performance goals of either of these options would contribute to the rapid development of the proposed SLGSM.

Conventional Trident Modification

CTM is the proposed U.S. Strategic Command (STRATCOM) near-term CPGS solution. It involves an adaptation of the Trident II (D5) missile for 2 of the 24 missile tubes on each SSBN. Each missile would nominally carry four modified Mk4 reentry vehicles.[5] The proposed deployment on the existing SSBN fleet allows the existing infrastructure of crews and training to be used, thus minimizing the costs of deploying the system and giving high confidence in its reliability.

The baseline Trident II (D5) is a stellar-navigated, inertially guided ballistic missile with a range of more than 4,000 nmi. Powered by a three-stage solid-rocket propulsion system, the vehicle achieves velocities in excess of 20,000 ft/sec. D5 missiles are carried in each of the 24 launch tubes on the 12 deployed Ohio-class SSBNs, which operate in both the Pacific and Atlantic Oceans. The Trident II (D5) has been deployed for more than 15 years, and extensive end-to-end system testing has shown the system to be extremely reliable, with 117 consecutive successful flight-tests as of this writing.

Overall the CTM is proposed as a modification of the baseline Trident system through (1) the replacement of the nuclear warheads with a conventional warhead consisting of dispersable KEPs on two missiles; (2) modification of the Mk4 reentry bodies through the addition of an aft extension (called a backpack) containing an aerodynamic control system, and a Global Positioning System (GPS) receiver and inertial navigation package to improve accuracy; and (3) modification of the launcher control system and its interface with the missile to allow positive control of mixed nuclear and conventional loads. The existing Trident warheads are unguided after they leave the third-stage "bus."

The KEP warhead is an area weapon capable of attacking soft targets whose position is known with accuracy on the order of meters. The warhead consists of many tungsten rods deployed by an explosive charge to create a relatively uniform dispersion of individual small rods. The technical issues associated with the KEP warhead are described below in the subsection "Weapons Effectiveness."

---

[5]The Navy uses the term "reentry body, or RB" as opposed to "reentry vehicle, or RV." This report uses both terms interchangeably.

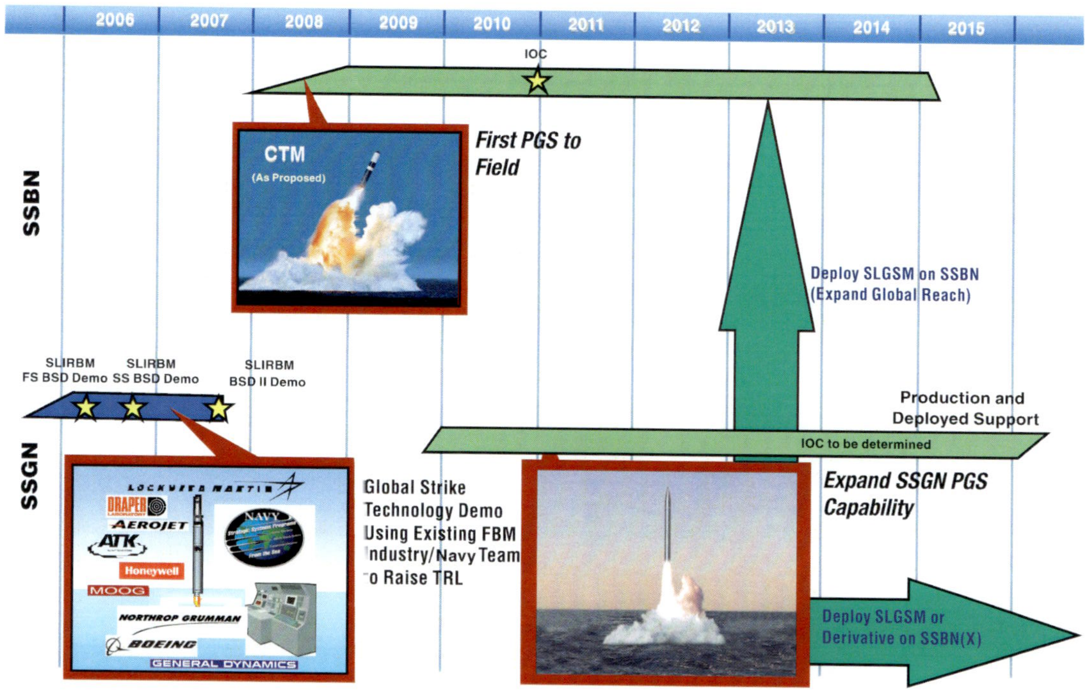

FIGURE 4-9 The prompt global strike (PGS) baseline system is the Conventional Trident Modification (CTM). The submarine-launched intermediate-range ballistic missile (SLIRBM) technology demonstrations in parallel with the CTM technology development provide the basis for subsequent development of the Submarine-Launched Global Strike Missile (SLGSM). NOTE: FS, First Stage; BSD, Booster System Demonstration; SS, Second Stage. Remaining acronyms are defined in Appendix A. SOURCE: Office of the Technical Director (Navy), Strategic Systems Programs, provided to the committee on December 18, 2007, Washington, D.C.

The key vehicle components are the modified reentry body and its integrated backpack for guidance and control. The primary issues in system development are navigation, guidance, and control, and weapon dispersal. The CTM proposal is backed by detailed design, analysis, and flight-testing. Non-nuclear reentry vehicles with the backpack have been flight-tested as described later in the subsection "Guidance, Navigation, and Control Accuracy Issues," and KEP payloads have been evaluated in sled tests and other short-range flights at White Sands Missile Range, New Mexico. The backpack contains a GPS-aided inertial guidance system and a four-flap aerodynamic control system (derived from the E2 test) for the maneuvering required to achieve the desired terminal accuracy. As discussed below in the subsection "Guidance, Navigation, and Control Accuracy Issues," testing to be carried out in the CTM development will provide crucial demonstrations of the achievable military utility of the ballistic delivery of conventional munitions. If CTM R&D is delayed, it may delay any other CPGS by years because of the need to do careful component-level and full-system-level testing on many more components.

Without special precautions, the proposed CTM program would raise the specter of an accidental launch of a nuclear weapon when the intent is to launch a conventional weapon because (1) both kinds of weapons would be carried on the same submarine platform, and (2) prompt global strikes may often allow little time for second checks. The technical issues associated with the nuclear surety of a mixed load were discussed in Chapter 2.

CTM with Modifications to Increase Payload Capabilities

Reentry Vehicle Modification for Vertical Attack There are significant advantages for the reentry vehicle to maneuver sufficiently at all ranges to be able to dive on the target vertically. This capability is needed to deny an adversary the option of positioning key assets in steep terrain or between structures that might shield them from weapons coming in at ballistic reentry angles, which typically are less than 30 degrees from the local horizontal for long-range ballistic flight. The Navy's current CTM proposal for initial conventional payloads does not offer this capability at the ranges of interest; however, the LETB test (see also the subsection "Guidance, Navigation, and Control Accuracy Issues") illustrated such a capability for a shorter-range application. The configuration tested was a biconic "bent-nose" variant of the baseline CTM payload, a larger variant of which is also the basis for the proposed SLGSM reentry vehicle. For CTM's longer ranges, this reentry payload would require a thicker heat shield and somewhat larger flaps than the version flight-tested most recently. The bent-nose biconic vehicle with the thicker heat shield is very similar to the Navy Strategic Systems Mk 500 evader maneuvering reentry body (see Figure 4-2), which was flight-tested numerous times on Trident I and on Minuteman missiles during the 1970s. It appears that this would be a minimum-risk development that would defeat an existing tactical countermeasure and that it ought to be evaluated for possible deployment as part of the initial CTM program. A primary difference between this proposed option and the SLGSM reentry vehicle is the size of the warhead that can be carried.

A Committee-Proposed Additional CPGS Option: The CTM-2 CTM as proposed is constrained to small conventional payloads primarily for missions against fixed soft targets. Another possibility discussed by the committee is a two-stage Trident configuration (CTM-2), a minimum-modification natural evolution from CTM, with payload volume and weight capacity offering a great deal of flexibility on what can be carried. A basic Trident missile with its third stage and two of its four post-boost control-gas generators removed (a simple modification that would not invalidate the D5's long test history) could carry a very large single- or multiple-weapon payload as long as it fits under the existing nose fairing or shroud. These modifications would expand the set of CPGS missions that could be effectively performed. This report refers to this two-stage Trident modification as CTM-2.

Figure 4-10 illustrates this flexibility, showing as examples that one 3,000 lb penetrator weapon can be carried by a CTM-2 missile to a ballistic range of about 4,000 nmi, and that a 1,500 lb penetrator weapon can be carried to more than

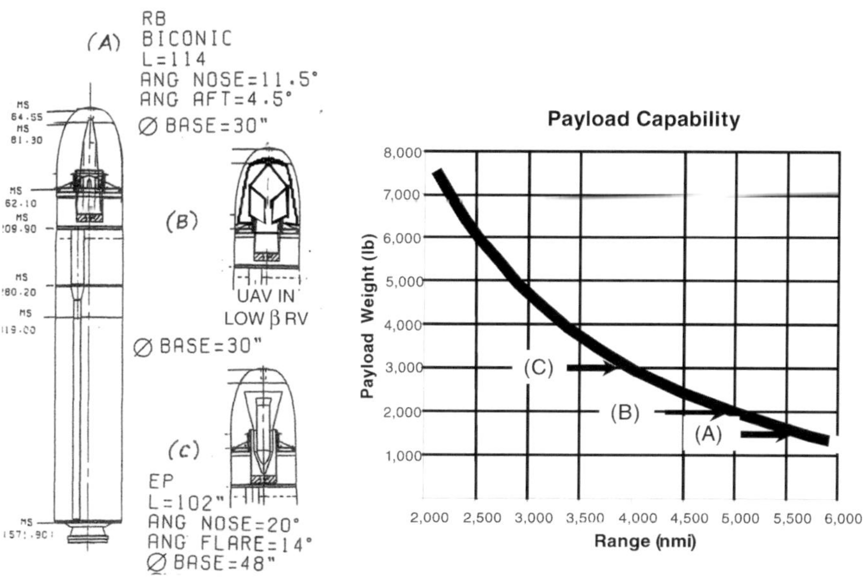

FIGURE 4-10 Candidate configurations for the committee-proposed Conventional Trident Modification (CTM)-2 illustrating concepts for integrating large payloads on a two-stage Trident missile, and results for range versus payload trade-offs for ballistic flight.

5,000 nmi.[6] In order to attack small, hardened buried targets, an earth-penetrator munition weighing on the order of 1,000 lb is required. Housed in a larger version of the biconic maneuvering RV just described, the total vehicle weighs about 1,500 lb. The maneuvering capability would not only allow this weapon to arrive on target vertically, but by choice of maneuver, the impact velocity could be controlled to be the roughly 0.5-1.0 km/sec that is best for penetrator effectiveness. The maneuver used to reduce velocity is typically a pull-up followed by a pull-down to vertical flight. Development issues for guidance, navigation, and control capabilities in this case are more severe than for CTM, as discussed below in the subsection "Guidance, Navigation, and Control Accuracy Issues."

For the biconic vehicle, several hundred additional miles in range and footprint are achieved by aerodynamic maneuvering to dump energy for best penetrator-munition performance. Four CTM KEP warheads also could be carried to the same range. On a longer development timescale, a ballistically delivered armed UAV with a half-hour endurance and its own search-and-target-acquisition capability could be developed with a gross weight of less than 3,000 lb, enabling mobile targets to be held at risk. This concept is discussed along with dispensed munitions later in this chapter.

While this CPGS option was not proposed by the Navy, the committee views the two-stage conventional Trident CTM-2 as an interesting evolutionary approach to CPGS because it may allow relatively early introduction (i.e., 2 years after CTM), and it has great payload flexibility and growth potential. It might or might not be different enough from the traditional three-stage Trident missile to satisfactorily address the ambiguity problem that some have been concerned about (see Chapter 3).

Submarine-Launched Global Strike Missile

For the longer-term CPGS role, the Navy proposes a small new, single-warhead, two-stage missile known as the Submarine-Launched Global Strike Missile, or SLGSM. The SLGSM is an intermediate-range ballistic missile that would be launched primarily from forward-deployed SSGN platforms and also, to increase geographic coverage, from SSBNs. The SSGNs are four Ohio-class submarines that were modified in ways that prevent carriage of the D5 missile. SSGN missions currently focus on Special Operations Forces and strike using Tomahawk missiles. The SLGSM concept involves the development of a new, two-stage solid-rocket propulsion system which, together with a reentry vehicle capable of a limited amount of gliding, is planned to deliver a 1,000-lb-class warhead to ranges in excess of 3,000 nmi. The missiles are sized such that two or

[6]Two 1,500 lb weapons could in principle be carried to 4,000 nmi but would require irreversible structural changes to the base frame of the missile, making this a less flexible option.

three SLGSMs could be held in a multiple all-up-round canister (MAC) within a single SSGN launch tube.

Once the technical solutions developed for the CTM have been tested, the more substantial investment needed to develop a more powerful CPGS system can be evaluated. A submarine-launched system allows global strike to be achieved with a smaller required delivery range and thus a less powerful ballistic launch motor. The correspondingly slower reentry speed results in a less-stressing lift into a glide path that allows long-range steering, as demonstrated for the LETB (see the subsection below titled "Guidance, Navigation, and Control Accuracy Issues") and envisioned for all the more-advanced delivery systems. Thus, the development path shown in Figure 4-9 for the SLGSM has strong feasibility of concept.

The specific plan for the SLGSM is to achieve greater military utility by providing a prompt strike capability from forward-based naval platforms, specifically SSGNs, allowing larger payloads compared to those of the CTM and a more flexible terminal trajectory through the use of a reentry vehicle with a moderate lift-to-drag ratio. The system requires the development of a new missile, including new solid-propellant booster motors and scaled derivatives of existing reentry vehicles. Rocket motor technology consistent with a shorter-range submarine-launched, intermediate-range ballistic missile (SLIRBM) has undergone two static firing tests (the SLIRBM demonstrations illustrated in Figure 4-9). The terminal maneuverability and accuracy will be effected with a medium-lift reentry vehicle, which is a scale-up of the LETB-tested bent-nose design, as illustrated in Figure 4-11, and is envisioned to provide the flexibility to contain different types of warheads. Experience from testing of the CTM reentry body will be important in the final development of SLGSM's RV.

For SLGSM's RVs, the challenges of pull-up into the glide path and thermal management during the glide are much less than those of the longer-range systems. Because similar boost stages and the payload have been tested for a similar concept, the committee considers that planning for SLGSM is a useful path in the evolution of technical capabilities beyond the CTM.

Range, Payload, and Size Trade-offs The committee has concerns about two aspects of the Navy's current concept for the SLGSM. First, it has a slenderness (length to diameter) ratio substantially greater than that of any existing underwater-launched vehicle. Second, it is planned to house two per SSGN tube, while the committee recommends that there be three per tube to increase firepower.

The Navy's current concept for SLGSM has a 38-inch-diameter missile with a slenderness ratio substantially greater than 12. The committee is concerned that a missile so long and slender may be subject to excessive bending stress as it leaves the tube of a submarine underway and enters the cross-flowing water. Shortening the missile would reduce its range for a given diameter. Figure 4-12 shows the results of independent missile simulations done by this committee, constraining the length-to-diameter ratio to 12. The figure shows payload versus range for

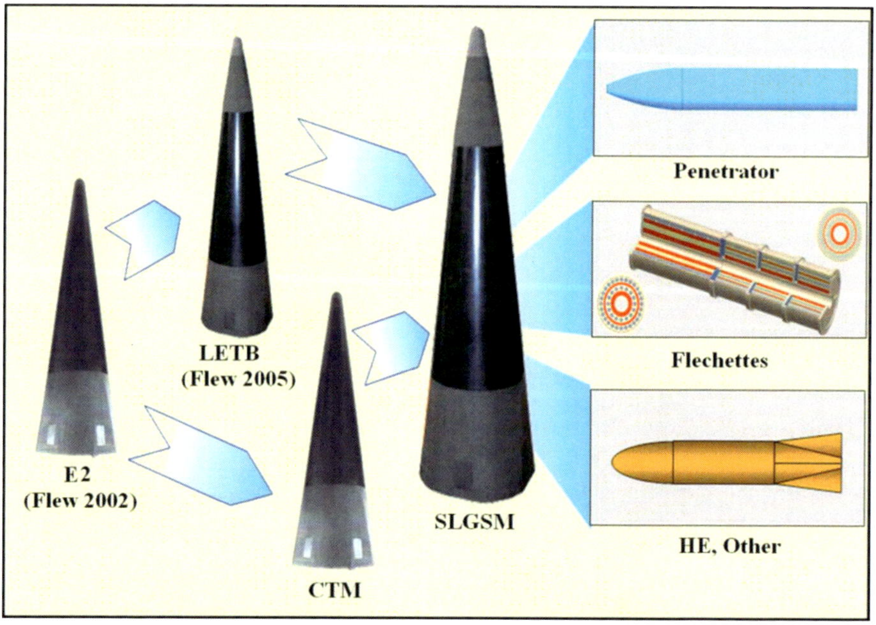

FIGURE 4-11 Relationship of the proposed Conventional Trident Modification (CTM) and Submarine-Launched Global Strike Missile (SLGSM) reentry bodies to the tested Enhanced Effectiveness (E2) Test Bed and Life Extension Test Bed (LETB) designs. The larger SLGSM would allow larger and more diverse payloads. SOURCE: Office of the Technical Director (Navy), Strategic Systems Programs, provided to the committee on December 18, 2007, Washington, D.C.

two- and three-stage missiles of different diameters and propellant classes. In its proposed two-stage configuration, when limited to length-to-diameter ratio of 12, a 38-inch-diameter SLGSM can carry a 1,500 lb payload to a ballistic range on the order of 2,000 nmi. The maneuvering biconic reentry vehicle can extend that range several hundred miles as it dissipates velocity. Since this range appears adequate for CPGS, the committee recommends that the Navy's Strategic Systems Programs (SSP) Office conduct a careful analysis of structural bending loads during launch before exceeding a slenderness ratio of 12 in the SLGSM design.

Shown in parentheses in Figure 4-12 is the number of missiles of each size that fit in a MAC in an SSGN tube. Three 38-inch-diameter SLGSM missiles should fit in each launch tube if the MAC for SLGSM is designed for the depth-charge shock levels currently used for Trident in SSBNs. (The MAC used in SSGNs with Tomahawk missiles is designed to a greater shock-mitigation requirement.) The committee believes that the 50 percent increase in firepower

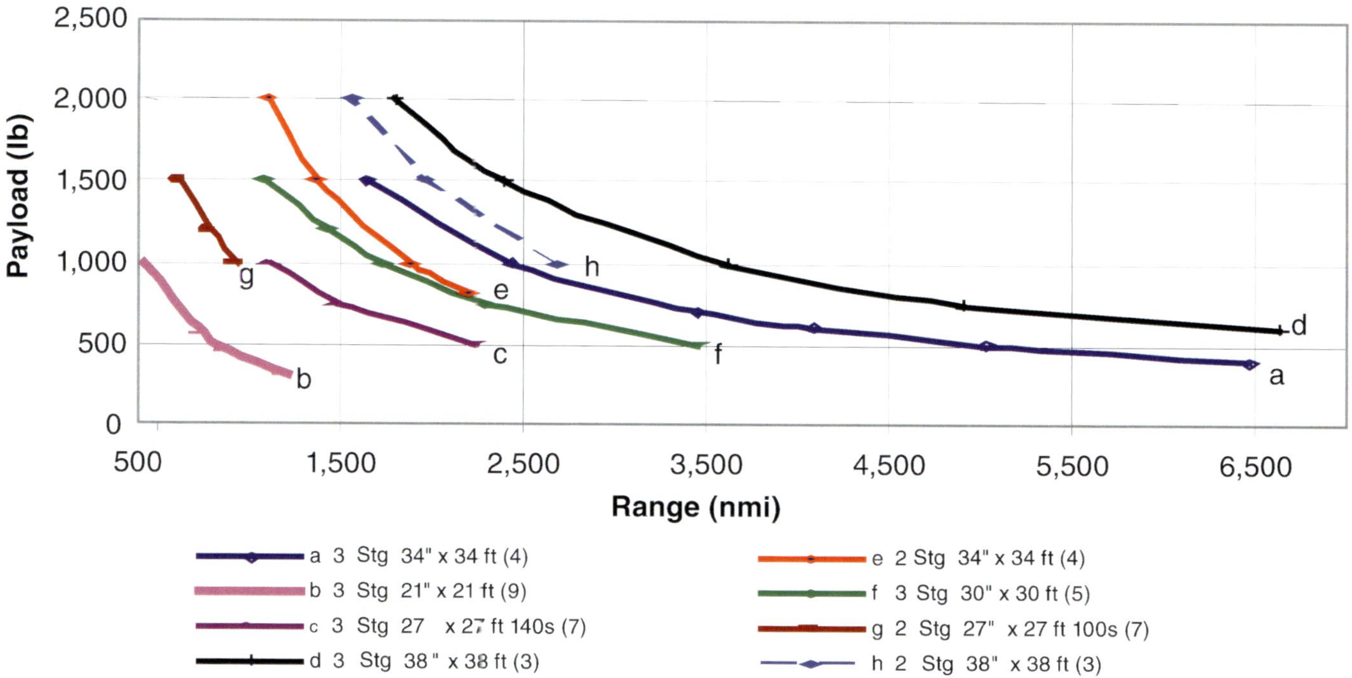

FIGURE 4-12 Payload versus ballistic range capability for notional two- and three-stage Submarine-Launched Global Strike Missile (SLGSM) configurations. All vehicles are constrained to a length-to-diameter ratio of 12. The number in parentheses indicates the number of missiles that can fit in a single launch tube of a nuclear-powered guided missile submarine.

is important enough to consider a reduction in the shock specifications, and the committee recommends that SSP look carefully at this issue.

The committee is concerned that the DOD's current plans for CPGS may be too sanguine in terms of its predictions of target geolocation accuracy, weapon delivery accuracy, and lethality of weapons. To some degree, these concerns could be addressed by carrying three missiles per tube and firing a patterned laydown. Figure 4-12 explores these possibilities. Figure 4-12 also shows that the same missiles with a third stage added can carry a smaller Mk 5 RV delivering about 700 lb of throw weight to a range of at least 5,500 nmi.

SLGSM design is driven by many factors that the committee cannot, of course, examine in adequate depth. It recommends that the Navy continue to explore the trade-offs that Figure 4-12 addresses.

Land-Based Missiles

The long history of U.S. nuclear-armed land-based ballistic missile programs provides a solid foundation for land-based CPGS concepts. These concepts can be divided into conventional ballistic missiles and boost-glide missiles, which can be CONUS-based or forward-deployed. This section discusses current Service programs, alternative concepts under active study, and other possible system concepts that come immediately to mind but are not being pursued, together with the reasons why.

Conventional Ballistic Missile: Conventional Minuteman

Just as the Navy has proposed to develop a Conventional Trident Modification, the U.S. Air Force could develop a Conventional Minuteman Modification (CMM). Warhead technology the Navy is developing for CTM could be applied as well to Minuteman, drawing on the several hundred boosters available in storage. Land-basing would avoid whatever additional risks to the nation's nuclear deterrence capability are engendered by mixed loads on SSBNs. Similar to CTM, this CMM concept would be effective against a limited class of targets and would be indistinguishable from its nuclear-armed progenitor externally and in-flight. In the committee's opinion, even if consensus were to emerge suddenly on the desirability of this CMM concept, it would probably take several years longer to field than would CTM because of real-world issues relating to the current-day status of programs, organizations, authorization delays, and other factors.

A new land-based conventionally armed ballistic missile could be developed with a more capable warhead, evolving later to even greater capabilities, as is the case for sea-launched concepts discussed above. However, it would have the basing and arms agreement issues discussed in Chapter 3. It also would have no significant performance advantage over a sea-launched missile, with the exception of potentially being able to deliver a larger warhead, which could increase effectiveness against some buried targets—although the relative ease with which facilities can be deeply buried makes this a losing game.

Gauging the balance of its advantages and disadvantages, the U.S. Air Force has decided not to pursue the development of a purely ballistic land-based conventionally armed ballistic missile, and the committee supports the Air Force's decision, as discussed in Appendix I.

Boost-Glide Missiles: Conventional Strike Missile and Advanced Hypersonic Weapon

A proposed alternative to a conventional ballistic missile is a (hypersonic) boost-glide vehicle—here a rocket is used to boost to high speed an aerodynamically controlled glide vehicle that maneuvers to the target. Concepts in which the initial phase of flight includes a ballistic segment have been proposed, while other concepts fly entirely endoatmospheric trajectories. Following apogee, the glide vehicle descends into the atmosphere where a pull-up maneuver is executed to position the vehicle on an equilibrium glide slope. The vehicle then glides unpowered to the target area. Both CONUS-based and forward-deployed boost-glide vehicles have been proposed. Compared with forward-deployed systems, prompt strike from within the CONUS requires a larger rocket and results in a flatter reentry angle and higher reentry speeds. The larger systems allow larger payloads than those that could be carried by CTM or SLGSM; however, the higher reentry speed renders more difficult the control, guidance, and navigation needed for accurate targeting, and the long exposure to high heat flux complicates thermal management. Nevertheless, developing a longer-range glide capability for the reentry vehicle in the form of a 161-inch boost-glide vehicle (larger than the modification proposed for the SLGSM) is the basis for the land-based options listed in Table 4-1. Additional details on the proposed boost-glide systems are provided below. The reader is also referred to Appendix G for additional details on the characteristics of the boost-glide trajectory.

CONUS-Based Boost-Glide Vehicle: Conventional Strike Missile The primary concept for a CONUS-based Conventional Strike Missile is based on a modified ballistic launch vehicle together with a scaled version of the advanced maneuvering reentry vehicle (AMaRV). The CSM concept uses a Minotaur III-class booster, a configuration using no longer operationally deployed Peacekeeper rocket motors. The reentry vehicle will have capabilities for pull-up and hypersonic glide similar to those proposed for SLGSM, but at much higher reentry speeds and for much longer times in the atmosphere.

The Air Force has programmed $477.7 million over the period FY 2008 to FY 2013 to develop and demonstrate CSM in three flight-tests, as illustrated in Figure 4-13. The weapon system is proposed to be based at Vandenberg Air Force Base, California, and Cape Canaveral Air Force Base, Florida (pending resolution of

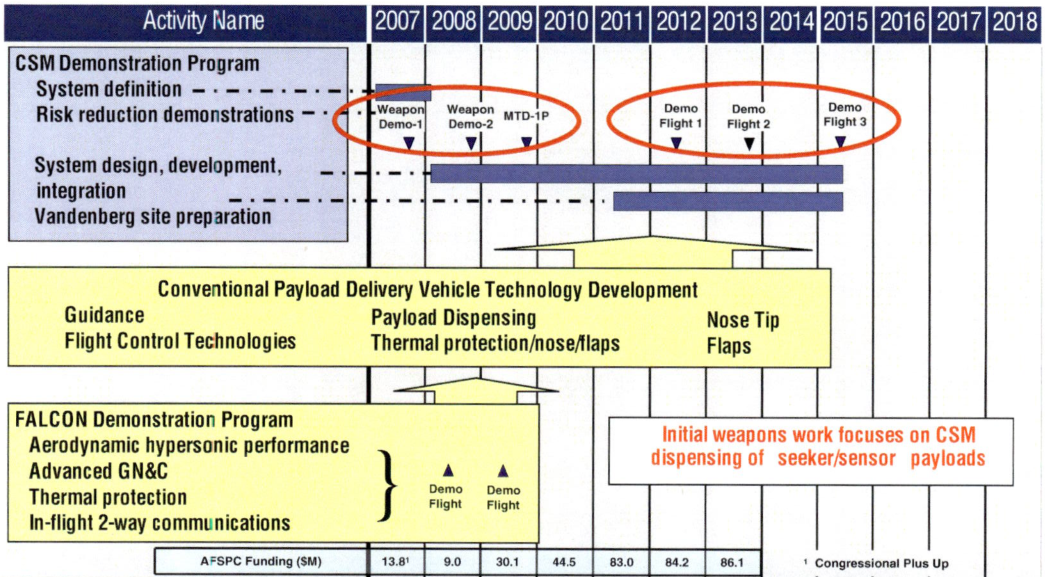

Demo	Location	Description
Weapon Demo -1	Eglin AFB	BLU-108 weapon software qualification
Weapon Demo -2	Eglin AFB	BLU-108 lethality demo
MTD-1P	White Sands	BLU-108 dispense/environment demo

Demo	Location	Description
Demo Flight 1	Vandenberg	Assess vehicle control/instrumented submunition
Demo Flight 2	Vandenberg	Demonstrate 800 second flight/BLU-108 dispense
Demo Flight 3	Vandenberg	Demo advanced thermal protection/global reach

FIGURE 4-13 The focus of the Conventional Strike Missile (CSM) Technology Demonstration Plan is on early weapons demonstrations to test the dispensing of existing off-the-shelf munitions, to address environmental testing and weapon requalification, with a priority given to "seeker/sensor"-type weapons that may address target location error and collateral damage issues. Ongoing research on thermal protection, guidance, navigation, and control, and in-flight communication will feed into later planned flight demonstrations. NOTE: Acronyms are defined in Appendix A. SOURCE: Maj. Steven Kravitsky, USAF, Headquarters, Air Force Space Command, "Prompt Global Strike Payload Development," presentation to the committee, May 11, 2007, Washington, D.C.

treaty issues), and to be capable of delivering KEP and penetrator warheads and many of the Service's air-launched weapons. The CSM's ability to alter its trajectory significantly would reduce overflight issues, enable significant cross-range diverting capability for in-flight retargeting, and permit tailoring the end-game approach angle for improved weapon system effectiveness.

The CSM concept pushes the performance envelope in the areas of thermal protection systems and air-launched weapon dispensing mechanisms. While a ballistic reentry body typically spends less than 60 seconds in the oxidative hypersonic environment, the first version of CSM is proposed to fly in it for about 800 seconds, stressing current technology. A planned second version of CSM would increase the maximum glide range to 9,000 nmi, which would require the development of new thermal protection technology to operate for up to 3,000 seconds in the stressing hypersonic environment. The CPGS AoA is modeling hypersonic boost-glide vehicles as slowing to Mach 5 to dispense their weapons. This speed seems at the same time aggressively high for dispensing and questionably low for surviving strong local air defenses. These technical risks are discussed in more detail below, along with a description of the Defense Advanced Research Projects Agency (DARPA) Force Application and Launch from CONUS (FALCON) Program that CSM is intended to leverage.

In the committee's view, the CSM effort planned and funded by the U.S. Air Force is optimistic for a program intended to result in a highly reliable, highly effective presidential-release weapon. In the committee's judgment, a prudently scheduled, well-funded program with adequate testing would have an initial operational capability (IOC) of about 2017 for an initial version of CSM (which is referred to in this report as CSM-1) that has an 800-second glide phase and is capable of delivering either KEP or penetrator warheads, and an IOC of about 2022 for a second version (referred to in this report as CSM-2) that is proposed to have a 3,000-second glide phase and to be capable of dispensing a wide variety of air-launched weapons at high speed or delivery of KEP or penetrator warheads.

Forward-Deployed Boost-Glide Missiles: Advanced Hypersonic Weapon Another design concept examined by the U.S. Army is a shorter-range but forward-based endoatmospheric hypersonic boost-glide vehicle, sometimes called the Advanced Hypersonic Weapon (AHW). It would have a range of approximately 4,200 nmi and could be based in Guam, at Diego Garcia in the Indian Ocean, and in Puerto Rico, with four missiles available at each launch site.

The Army Space and Missile Defense Command in Huntsville, Alabama, received additional congressional funding of $1.6 million in FY 2006 and $8.9 million in FY 2007 to do technology studies to support a similar weapon concept, but the Army has not programmed substantial out-year funds for serious weapon system development. Army officials informed the committee that technology developed in this effort would flow into the Air Force's CSM program.[7]

Hypersonic Cruise Missiles

In addition to ballistic and boost-glide missiles, hypersonic cruise missiles launched from submarines, surface ships, or aircraft potentially offer capability for the CPGS mission. In concept, these systems consist of two stages: the first-stage rocket-powered and the second stage powered by an air-breathing ramjet or scramjet (i.e., supersonic combustion ramjet) engine.

The advantage of the cruise missile is that the air-breathing engine uses oxygen from the atmosphere instead of carrying an oxidizer, allowing for a missile capable of large speed variation and diverting maneuvers, flexible approach to the target area, and the ability to control the trajectory for the deployment of submunitions, with potential for loitering and searching for mobile targets. One disadvantage is the relative immaturity of the air-breathing propulsion technology for missiles operating at speeds above Mach 4. Propulsion technology issues and readiness levels are discussed more completely below.

At present, cruise missiles such as the Tomahawk and conventional air-launched cruise missile (CALCM) are limited to subsonic speeds, with ranges of approximately 1,000 to 2,000 nmi (see the data presented in Figure 4-8). Figure 4-14 shows the results of independent calculations done by the committee on the characteristics of long-range subsonic and hypersonic cruise missiles. The potential performance variation is bounded on the lower end with an assumption that aerodynamic and propulsion efficiency remains constant with increasing missile size and on the upper end by an assumption that aerodynamic and propulsion efficiencies will increase with increasing missile size. For the lower performance numbers, the baseline engine and aerodynamic performance are constant with increasing missile diameter and consistent with those obtained from flight and laboratory tests for short-range tactical missiles. The upper curves account for potential performance improvements that can be realized for larger-scale flight vehicles. In these curves, the assumed improvement in aerodynamic efficiency is a 67 percent increase in the vehicle lift-to-drag ratio (increasing from 12 to 20 for subsonic missiles and from 3 to 5 for hypersonic missiles) over the range of missile diameters considered. This increase is consistent with the incorporation of larger, more efficient lifting surfaces on the larger missile. The assumed improvement in propulsion efficiency is a 33 percent improvement in engine specific impulse over the range of missile diameters considered. This improvement consists of an assumed increase from 2,656 seconds to 3,600 seconds for

---

[7]Col Paul Gydesen, USAF, Chief, Deterrence and Strike Division, Plans and Requirements Directorate, U.S. Air Force Space Command, discussion with the committee, February 23, 2007, Washington, D.C.

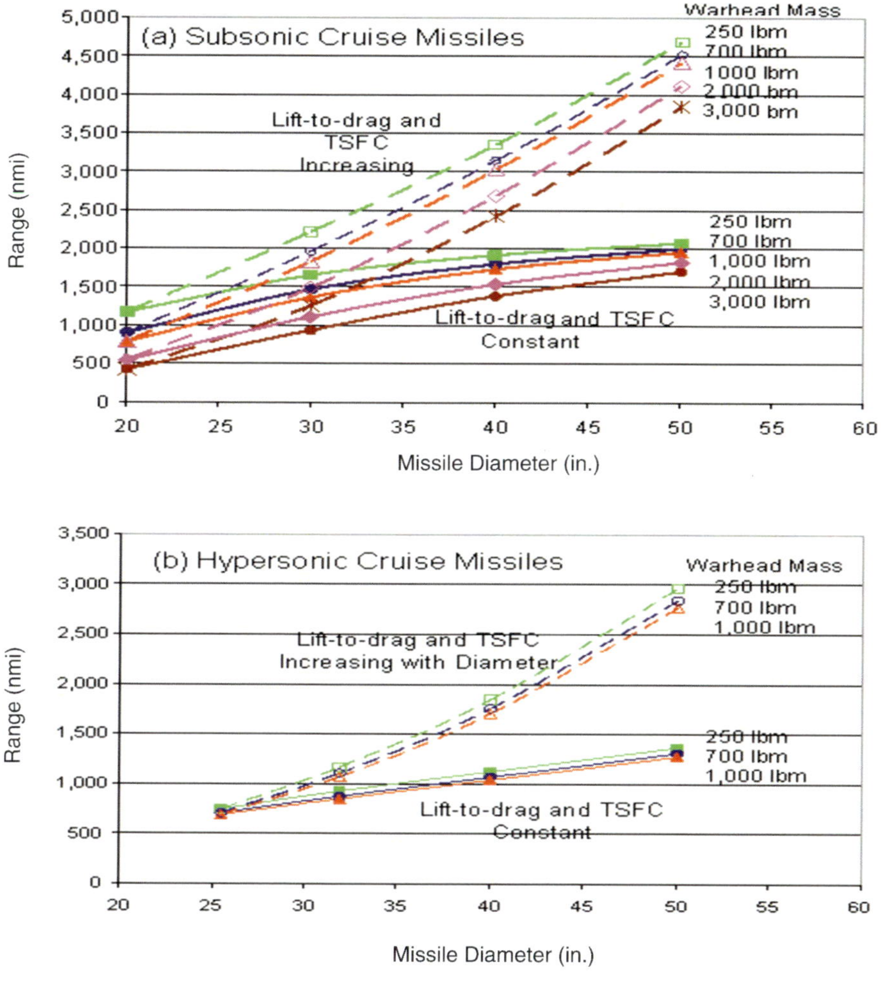

FIGURE 4-14 The impact of missile diameter and payload on range of notional (a) subsonic and (b) hypersonic cruise missiles. The solid curves are based on flight- or laboratory-demonstrated aerodynamic and propulsion technology. The dashed curves illustrate the potential performance improvements that result from potential aerodynamic and propulsion efficiency gains. NOTE: TFSC, thrust specific fuel consumption.

subsonic missiles. For hypersonic missiles, the engine specific impulse varies with speed, and the 33 percent improvement in specific impulse was applied as a scale factor across the speed range to account for the anticipated performance improvement associated with larger engines. When designed to constraints of the Vertical Launch System carried on surface ships, a hypersonic cruise missile would have a maximum range of approximately 1,000 nmi if operated at a cruise speed of Mach 4. While this range is much shorter than that envisioned for a CPGS system, this capability could become an important element of a future tactical system if routinely installed on forward-deployed ships and submarines.

The possibility of employing hypersonic cruise missiles from the SSGN platform also provides an interesting potential. In designing the cruise missile for integration into an SSGN canister, the missile cross-sectional shape will likely not be circular, but rather designed to fit within a pie-segment of the cylindrical launcher. In this case, a larger effective missile diameter can be achieved. If the SSGN launch tubes are modified to hold three hypersonic cruise missiles, calculations indicate that the powered range of the missile would be greater than 2,000 nmi (i.e., roughly the same range capability of a submarine-launched, intermediate-range ballistic missile).

Air-launched, hypersonic cruise missiles have been proposed in order to satisfy the demands of the CPGS mission (see Mach 6 Missile in Table 4-1). Launched from a bomber, the cruise missile would be a two-stage system consisting of a solid-rocket-powered first stage and a second-stage air-breathing cruise vehicle powered by either a ramjet or a scramjet engine. The air-launched missile would be capable of carrying a 2,000 lb warhead to a range of approximately 2,000 nmi. The air-breathing propulsion system would offer advantages in terms of trajectory flexibility and energy management, allowing for in-flight rerouting, avoiding overflight, and optimizing the terminal approach geometry.

Thermal management issues associated with hypersonic cruise missiles are quite different from those involving ballistic or boost-glide systems. Operating at a maximum speed of Mach 6, the heat transfer to the external air vehicle is substantially lower than that of reentry vehicles, and metallic structures can be used. In the internal portion of the flow-path, either high-temperature ceramic-matrix composites or fuel-cooled metallic structures can be used. Thus, many of the issues associated with ablative thermal protection systems, which are required for ballistic and boost-glide systems, are avoided.

RESEARCH AND DEVELOPMENT ISSUES

To varying degrees, the CPGS concepts described above share challenges associated with reliably operating in an extremely stressing and unforgiving environment. In the following subsections, several of the most important technology issues are described and discussed. But first a few comments on operating environment are needed. The first concerns the characteristics of ballistic reentry. One of the principal design factors is called the ballistic coefficient, β, which is defined as follows:

$$\beta = \frac{W}{C_D A}$$

where W is the weight of the reentry vehicle, C_D is the drag coefficient, and A is the cross-sectional area. Manned reentry vehicles typically operate with a low β, such that the reentry vehicle slows significantly in the upper portion of the atmosphere where the blunt vehicle shapes produce strong shock waves with significant energy dissipation through heating of the atmosphere, which minimizes the overall thermal challenge to the vehicle. This type of reentry was used in the Apollo program and the early photo reconnaissance missions. Low-β reentry could be useful in a CPGS mission context for concepts that deploy munitions or ISR assets, as discussed earlier.

Ballistic missile reentry vehicles are typically designed with a high β so that the reentry vehicle plunges into the deep atmosphere at high speeds. This makes defense very difficult (which was an early consideration in the development of ballistic missiles) and enables high accuracies with unguided RVs because of the short atmospheric transit time and resulting small drift in the inertial navigation system. The high-β reentry leads to an increased heat transfer, and thus the requirement for a robust thermal protection system.

A CONUS-based ballistic missile with a range of 7,500 nmi has a reentry speed of approximately 24,640 ft/sec. A forward-deployed ballistic missile with a range of 2,500 nmi has a reentry speed of approximately 18,300 ft/sec. While this velocity difference may not appear overly large, one must note that the kinetic energy is proportional to the velocity squared, so that the shorter-range reentry vehicle enters the atmosphere with a kinetic energy that is approximately 45 percent lower than that of the long-range system. This difference in entry kinetic energy results in significant differences in the stress placed on the thermal protection and guidance, navigation, and control systems.

Thermal Protection System

Conventional Approaches

Existing high-β ballistic reentry vehicles are designed with thermal protection systems (TPSs) that use ablative materials to survive an extremely high-heat-flux environment for the short reentry times. Modern reentry vehicles use a shape-stable carbon material nosetip with a carbon-phenolic heat shield for thermal protection downstream of the nosetip. Through extensive development and testing, carbon-phenolic material has proven to be the best choice based on its density, ablation characteristics, and relatively low thermal conductivity.

The CTM, SLGSM, and the initial variant of CSM are all based on existing TPS technology and existing thermal design and analysis techniques. Only CTM uses the existing TPS materials in the manner for which they were originally developed (i.e., in a ballistic reentry mode with minimal endoatmospheric maneuvering). CSM-1, and to a much lesser degree SLGSM, uses existing TPS technology in a boost-glide trajectory in which the TPS is exposed to the environment for longer time periods than for conventional ballistic RVs.

As discussed above, SLGSM is an intermediate-range missile that delivers its RV on a conventional ballistic trajectory. Its Maneuverable RV (MaRV) is similar to the Mk 500 MaRV, which was developed and tested at much longer ranges for glides, error correction, and desired terminal conditions. While it therefore must dissipate less heat input and has a shorter soak time (atmospheric flight) than long-range CONUS-based concepts, the soak time effects must be tested. (If that payload were also used for CTM-2, it would need to be capable of handling total heating and soak times closer to the CONUS-based CSM-1 requirement if a gliding trajectory was flown.) This longer exposure time introduces uncertainties in the TPS performance due to the potential for nonlinear coupling of the body shape changes with aerothermodynamics, especially with respect to uncertainty in the prediction of aerodynamic boundary layer transition. Application of existing TPS technology to SLGSM and CSM-1 concepts will need to be fully evaluated as part of the RDT&E associated with these systems.

Thermal protection issues must also be addressed in designing for weapons effectiveness. Penetrating warheads are limited by structural considerations to impact speeds of approximately 1 km/sec. Thus, ballistic and boost-glide delivery systems must be operated in such a manner that the kinetic energy of the warhead is dissipated sufficiently prior to impact. For the dispersal of dispensed munitions or UAVs, even slower final speeds may be necessary. For CONUS-based global range systems that reenter at 7.9 km/sec, 30.7 MJ/kg of kinetic energy must be dissipated if the impact speed is limited to 1 km/sec. For high-β reentry vehicles, the slender aerodynamic shapes result in minimal heating of the atmosphere, with most of this dissipated energy entering the vehicle prior to being released as radiation cooling or ablation products. For a shorter-range SLGSM or forward-deployed land-based boost-glide system with an initial glide velocity of approximately 6 km/sec, the kinetic energy to be dissipated is approximately 17.5 MJ/kg.

Advanced Thermal Protection System Concepts

The development of a capability for extremely long range glide vehicles, as envisioned in the advanced CSM-2 concept, will require the development of a novel thermal protection system capable of operating at hypersonic speed in the atmosphere for times up to 3,000 seconds. With the RV based on carbon-carbon (C/C) material rather than carbon-phenolic material, the aim is to develop an integrated TPS system with a shape-stable nosetip, an acceptable ablation rate over the flight trajectory, and minimal heat transfer to the internal portion of the reentry vehicle.

Numerous technical challenges exist with the development of this type of thermal protection system, including the development of techniques for the accurate prediction of aerothermodynamic loads and ablation rates, manufacturing processes for the production of large-scale C/C aerostructures, and insulation for the protection of internal components.

The challenges associated with the development of a 3,000-second TPS system have been undertaken by the DARPA FALCON Program. In this effort, two hypersonic test vehicles (HTVs) will be flown between Vandenberg Air Force Base and Kwajalein Atoll (location of the U.S. Army's Reagan Test Site) to demonstrate the performance of a long-duration TPS (as well as demonstrating guidance, navigation, and control accuracy and in-flight communications). The flights are scheduled for 2009 and 2010, so early information on the performance of this new class of TPS will be available to support decisions on the development of long-range systems.

Guidance, Navigation, and Control Accuracy Issues

As noted earlier, ballistic missile delivery systems were developed for nuclear weapons delivery for which, owing to the large damage area of the weapon, achieving the required accuracy in the placement of the weapon is relatively easy to accomplish. In contrast, for conventional weapons, accuracy of placement (technically referred to in terms of the circular error probable [CEP] or spherical error probable [SEP]) is essential in order to obtain the desired effects on the target. Obtaining this increased delivery accuracy is a key technological challenge in developing a conventional prompt global strike capability.

Obtaining high delivery accuracy for a ballistic missile requires that the reentry vehicle be steerable. Because the vehicle is moving extremely rapidly and thus spends a relatively short time within the atmosphere, controlled steering to the desired target is technically challenging. Preliminary design solutions have been field-tested, specifically using modified Mk4 reentry bodies, and serve as the "proof of principle" for the proposals for ballistic missile delivery in CPGS. The two field-tests were the submarine-launched E2 Test Bed and the LETB (see Figure 4-11).

The E2 evaluation was conducted in October 2002 using a D5 missile. The modified Mk4 reentry body had an added flap actuator system for control and a GPS-aided Inertial Navigation System (INS). During a substantial part of reentry, GPS reception was lost owing to a plasma-induced blackout. Nevertheless, the flaps deployed and operated as predicted to provide three-axis flight control (roll, yaw, and pitch). Under flap control, the reentry body (RB; see footnote 5, above) was able to turn to its target, and guidance and control were sufficient to steer the RB to the target within several meters of the onboard navigation solution. In other words, the RB came within several meters of where the navigation system calculated that the target was located. There was, however, a large navigational error, so that the actual impact point was well off the target. The correction for the navigational error is expected to be technically feasible. Design analysis has been performed which indicates that a properly integrated GPS/INS, incorporating accurate alignment initialization (and, probably, adaptive control algorithms), will make it possible to reduce navigational errors to GPS-like accuracy.

The LETB modification to the Mk4 included, in addition to the flaps, a 2 degree offset biconic nose for extended range and added flight stability and maneuverability. The navigation system was similar to that flown in E2. During the terminal phase of the trajectory, the RB executed a pitch-up maneuver, diverting the RB an additional 25,000 feet downrange from the ballistic impact point and turning to its selected target. A 3-second loss of GPS reception of most satellites was attributed to a failure to maintain lock on the carrier during the high-*g* pitch-up maneuver. As discussed in the following subsection, it is believed that this problem can be addressed by using newer technologies able to withstand the effects of maneuvers up to 40 *g*'s.

The proposed CPGS ballistic delivery systems build on these earlier results, as illustrated in Figure 4-2. As has been discussed, the simplest proposal is the CTM, which builds on the E2 results to allow fine control of targeting around the normal ballistic reentry trajectory. The longer-term proposals would include extended reentry trajectories enabled by the adaptation of the biconic nose design used in the LETB. Thus, resolving the issues regarding guidance, navigation, and control that were demonstrated in the E2 and LETB tests remains a crucial step in establishing the feasibility and military utility of the proposed systems for CPGS.

GPS/INS Navigation

To achieve weapons effectiveness in all of the intended use-cases, strike accuracy within a few meters in each dimension is needed. Reducing miss distance to this level is a sizable challenge given the numerous sources of error (target location, guidance, navigation, and control) that contribute to it. Here the navigation error is addressed, that is, the terms in the error budget that relate to the knowledge of position, velocity, acceleration, and attitude of a delivery vehicle under navigation with a combined GPS/INS.

Both GPS and INS have strengths and weaknesses. The main advantage of GPS is that it provides positional data with considerable accuracy and bounded errors; however, it is susceptible to loss of signal. GPS uses a low-power signal that makes it fairly easy to compromise reception. Plasma formation, loss of carrier lock during maneuvers, and jamming are each capable of compromising reception for CPGS applications. An INS offers precision navigational data (acceleration, velocity, position, and attitude) in real time based on the gyroscopic inertial measurement unit (IMU), and it is far more resistant to denied service than GPS is. Thus, IMU can provide precision navigation in the vicinity of the ground target in the presence of GPS jamming or GPS blackout due to plasma sheath attenuation. An INS tends to accumulate errors over time, but feedback of accurate position from GPS can provide the needed corrections. Thus, the respective strengths of GPS and INS are complementary. When combined into a tightly integrated GPS/INS system, the shortcomings of either component operating individually are better managed.

Early versions of GPS/INS have suffered from susceptibility to loss of lock on the carrier loops during large-*g* maneuvers. However, integrated systems are now guaranteeing service at up to 40 *g*'s. Within this limitation, which is reported to be manageable for CTM,[8] present GPS/INS systems, properly implemented, are sufficient to guarantee navigational accuracy currently measured in meters, with continuing improvements in capability likely to be available.[9]

For the larger reentry vehicles, faster reentry speeds, or more stressing environment of highly maneuverable vehicles, the issues of maintaining GPS/INS lock will have to be addressed again. Similar issues will also arise for energy dissipation maneuvers prior to the release of penetrating warheads, dispensed munitions, or UAVs (see below). Energy dissipation maneuvers can be performed by the reentry vehicles, but acceleration-sensitive drifts in the IMU may impact accuracy. A complete assessment of the accuracy associated with such maneuvers will be required on a case-by-case basis.

Guidance and Control

Given the likelihood that GPS/INS solutions are sufficient for the requirements of CTM, the precision and accuracy of the aerodynamic response of the RV to guidance and control information are crucial to achieving delivery accuracy. Engineering analyses of proposed design for the CTM reentry vehicle have been performed; they indicate that achieving the control necessary for the required few-meter accuracy is feasible, if all systems perform at their expected technical capability.[10] Thus it is essential that flight-tests planned for CTM be carried out and that they be designed to demonstrate RV-delivery accuracy (e.g., relative to a specific target of known absolute GPS coordinates) and the factors contributing to the miss distances. The results from these tests will serve as essential information for design decisions in the more-advanced proposed systems concepts.

[8]CAPT Terry J. Benedict, USN, Technical Director, U.S. Navy Strategic Systems Programs, "CTM Brief to NAS (U)," presentation to the committee, February 23, 2007, Washington, D.C. (classified).

[9]"JASON Study for STRATCOM on Prompt Global Conventional Strike," JSR-05-450, July 29, 2005, presentation to the Defense Science Board, January 2007, Washington, D.C. (This document contains other restricted/controlled unclassified information as described in 5 U.S.C. 552(b).)

[10]CAPT Terry J. Benedict, USN, Technical Director, U.S. Navy Strategic Systems Programs, "CTM Brief to NAS (U)," presentation to the committee, February 23, 2007, Washington, D.C. (classified).

Munitions and Sensor Deployment

The technical demands on guidance, navigation, and control for high terminal accuracy of the reentry vehicle could be dramatically reduced if the reentry vehicle dispensed a secondary, steerable weapons delivery body. Several CPGS system concepts rely on dispensing submunitions, weapons, or UAVs; the dispensing may occur at either high-speed or low-speed conditions, following reentry, hypersonic glide, or cruise. High-speed dispensing of submunitions presents significant challenges regarding control of the dispensing munition to avoid recontact with the RV, while simultaneously enabling aerodynamic capture of the munition. If the dispensing occurs at hypersonic speeds, additional thermal protection will likely be required for the munition.

As an alternative to a high-speed dispensing, the reentry vehicle may slow to low speeds, at which the challenges associated with dispensing munitions have largely been solved. Slowing a high-β RV to low supersonic speeds presents challenges associated with dissipating the large-vehicle kinetic energy while simultaneously maintaining accuracy in the dispensing position, as discussed earlier. By contrast, the use of a low-β vehicle (as proposed in the DARPA Rapid Eye program) to slow to acceptable speeds is straightforward.

An example of a promising dispensing system is the UAV envisioned for CTM-2, which would have ballistic delivery with a very low-β reentry vehicle. A blunt reentry heat shield is proposed to slow the payload down to a velocity at which a ballute or drogue chute can deploy (á la Apollo or early film-recovery capsules). Then when the vehicle slows sufficiently, the proposed design is for the drogue to deploy the main parachute, slowing the system to a velocity at which it can safely extract the UAV from the back of the reentry shell, and the UAV can then search for and acquire the target with its onboard sensors and, upon verification and authorization, can engage the target.

This type of concept greatly reduces the need for accuracy in the navigation of the reentry vehicle, but challenges exist in the development of the extraction, erection, and control of the UAV. DARPA recently solicited industry for demonstration of a similar concept to rapidly deliver a long-endurance ISR UAV.

Propulsion System Development

Rocket Propulsion

New ballistic missile or boost-glide vehicles for the CPGS mission will require new solid-propellant booster motors. These motors will generally be classified as either Class 1.1 or 1.3 explosives, depending on the sensitivity of the propellants. Class 1.1 propellants are highly energetic propellants that may detonate under stressing conditions. Class 1.3 propellants, while not as energetic, will not detonate. Whether a propellant is designated Class 1.1 or 1.3 defines requirements for handling, storage, and so on.

The DOD prefers, for good reasons, that all new weapon systems incorporate insensitive (nondetonating) munitions, which requires a booster propellant to be Class 1.3. Since Class 1.1 propellants offer better performance, system developers, facing volume constraints, often "buy into" the handling issues of Class 1.1 propellants. The committee recommends that the potential advantages of any new system incorporating Class 1.1 propellants be evaluated in detail to ensure that the risk and complexity of using Class 1.1 propellants are warranted.

Air-Breathing Propulsion

Hypersonic cruise missiles require the development of air-breathing engines such as ramjets or scramjets. Ramjet engines are capable of powering the missile to speeds up to approximately Mach 4. Ramjets were first flown at speeds of Mach 4 in the early 1960s. The technology basis for ramjet propulsion systems is very mature, and ramjet-powered systems are currently operational in many countries.

The development of scramjet engines will be required to power air-breathing cruise missiles operating at speeds above Mach 5. This engine technology has been investigated extensively in a laboratory environment since the late 1950s, but the transition to flight demonstrations only began with the Kholod scramjet demonstration test-flights conducted between 1991 and 1998 (in concert with France and NASA), the Australian HySHOT II scramjet test in 2002, and the NASA X-43A Mach 7 and 10 flight demonstrations conducted in 2003 and 2004, respectively. Continuing technology demonstrations include the DARPA/Office of Naval Research (ONR)-funded Hypersonic Flight Demonstration (HyFLY) and USAF/DARPA-funded X-51 programs aimed at demonstrating technologies for Mach 6+ cruise-missile applications.

In-Flight Communications

Communicating to a CPGS weapon in the midcourse or terminal phases of its trajectory offers significant operational advantages for all proposed CPGS options. This subsection discusses those advantages and candidate means of accomplishing the communication.

Applications

For all CPGS options, in-flight communication is needed to support flight aborting and battle damage assessment (BDA). Owing to its importance, BDA is discussed separately in a later subsection. For boost-glide missiles and hypersonic cruise missiles, in-flight communications would enable in-flight targeting update and target reacquisition and verification.

In-Flight Targeting Update As discussed in earlier reports of the Naval Studies Board[11] in-flight targeting update is an essential part of any long-range system for hitting moving ground targets. This is because it enables an appropriate balance of the burden of performance between the off-board targeting system and its geolocation accuracy and report frequency on the one hand, and the weapon and its seeker's ability to detect and classify on the other. As a practical matter, boost-glide missiles or hypersonic cruise missiles must be given a steady stream of reports in order to intercept the *right* moving target. Brought into the vicinity of the right target by the in-flight targeting updates, the boost-glide missile or hypersonic cruise missile can dispense a seeker-guided weapon (subject to the technical constraints of real systems as discussed below in the subsection "Mobile Targets") for target kill. Commanders may also want or need the capability to terminate a mission after launch.

Retargeting and Loitering Capability It is possible that a more valuable target may emerge after the launch of a CPGS weapon, or that a primary target may be lost (e.g., may have gone underground), and as a result a secondary target becomes more desirable. In-flight communication would give the commander flexibility to make the changes necessary in these cases.

Boost-glide missiles and hypersonic cruise missiles have the significant advantage of being able to maneuver to updated target coordinates following a launch. In-flight communication can exploit this capability by giving the weapon a new pop-up target that had not been detected prior to launch. Because of the expense of the weapon, there must be a sufficiently valuable target available as a default if no pop-up target arises.

The committee believes that retargeting capability would be of relatively little value in a presidential-release CPGS weapon, because it is intended to be used against only one or a few high-value targets in a given area.

Communications Means

The most promising means of communicating with a CPGS weapon in-flight is through existing ultrahigh-frequency (UHF) satellites. This frequency band permits the weapon to employ a simple, omnidirectional antenna. At UHF, the transmission data rate is limited, but achievable rates would accommodate all of the

[11] Naval Studies Board, National Research Council, 2000, *Network-Centric Naval Forces: A Transition Strategy for Enhancing Operational Capabilities,* National Academy Press, Washington, D.C.; Naval Studies Board, National Research Council, 1993, *Space Support to Naval Tactical Operations* (U), National Academy Press, Washington, D.C. (classified).

applications discussed here, including battle damage assessment. With sufficient priority (and a presidential-release weapon should have it), satellite channels can be made available reliably. The Tactical Tomahawk has such a link. The potential for establishing such capabilities for CPGS may be influenced by the plasma sheath that can form during reentry, interfering with radio-frequency communications channels (see the subsection "GPS/INS Navigation" above). The formation of a plasma sheath is highly dependent on the specific characteristics of the reentry vehicle. Vehicle speeds above Mach 10 may result in a plasma sheath, while speeds about Mach 20 virtually guarantee it. Altitude is also a factor, with plasma effects occurring between approximately 30,000 feet and 300,000 feet. The shape of the reentry body is important, with blunt bodies causing higher-density plasma. Finally, ablative materials and materials high in plasma-inducing contaminants are likely to increase plasma density. Thus, careful design and testing to evaluate the impact of plasma formation around the reentry vehicle will be important for in-flight communication, as well as for GPS reception, as discussed above.

Weapons Effectiveness

One of the warheads proposed for employment in some proposed CPGS systems is a cluster of segmented tungsten rods that are explosively deployed through the heat shield of the RV at a time determined by a fuze. The result is a selectable pattern of the rod segments or KEPs that are dispersed over the desired target area, impacting at a velocity of more than 5,000 ft/sec. If the fuze is set not to fire, the resulting densely packed cluster of rods serves as a 250 lb unitary slug (penetrator) which, along with a RV structure, impacts with kinetic energy sufficient potentially to penetrate a multistory building or to create a crater several meters across and deep. The issues of the effectiveness of such kinetic energy weapons against various types of targets are discussed below.

Fixed Soft Targets

Most fixed soft targets are susceptible to attack by unitary blast-fragmentation or kinetic energy warheads. As previously noted, the initial CTM would carry up to four reentry vehicles equipped with advanced navigation, guidance, and control capabilities. Each reentry vehicle would carry a warhead consisting of dispersible KEPs. Using approximately 1,000 small tungsten rods deployed by explosive charge, a relatively uniform pattern of small KEPs is created. The kinetic energy of a single rod is approximately the same as that of a 50-caliber bullet. The dispersion radius of the pattern can be varied by varying the height at which the warhead is triggered, with constraints imposed by atmospheric drag on the dispensed rods.

If a completely uniform pattern could be achieved and a single rod placed in each square meter on the ground, the dispersion pattern would have a diameter of 35.7 meters. If the target is such that a denser pattern is required with, say, 16 rods per square meter, a 1,000-rod warhead would be capable of producing a dispersion diameter of 8.9 meters. In both cases, the dispersion diameter is greater than the expected miss distance, so the KEP warhead is anticipated to be effective against soft targets. If the target location error was somewhat larger and a dense pattern was needed over a larger area, the multiple RV capability offers the option of patterning entirely on the single target.

In understanding the military effects of KEP weapons,[12] it is important to realize that there is no explosive blast (other than that used to disperse the projectiles), and thus the extended damage due to overpressure does not occur. Instead, direct structural damage is dependent on the materials response of the target. Many structural elements, such as the wall of a vehicle, the face of a radar dish, or the roof a building, will experience local damage directly at the point of impact, comparable to the type of limited damage caused by meteor strikes, as shown in Figure 4-15. While an increased speed of the projectile increases its kinetic energy, the efficiency with which the energy is transferred to cause lateral damage to the target generally does not increase correspondingly. The size of the damage region will be comparable to the size of the projectile, as shown in Figure 4-15. In the case of the small tungsten rod fragments of a KEP, the structural damage area per particle will be much smaller.

Thus, estimating the actual military weapons effectiveness has been the topic of analysis in terms of the susceptibility of the target to the specific action of the small kinetic energy particles of the CPGS warhead.[13] In addition, the types of damage must be ranked in terms of how long lasting the damage is. Effectiveness rankings can include an attack that merely delays enemy action for a few seconds or minutes, or an attack that requires the enemy to delay action until a repair can be effected. The timescale required for the repair then becomes a further criterion for ranking the effectiveness. Analyses of a wide range of target types under consideration for CPGS have been performed. The results indicate that in most cases, a single CTM KEP will have a high kill probability against fixed soft targets if target geolocation accuracy and guidance, navigation, and control accuracy are as predicted. Current plans call for high-speed sled tests of the KEP warhead and for continued modeling of the effectiveness of the KEP warhead against classes

[12]CAPT Terry J. Benedict, USN, Technical Director, U.S. Navy Strategic Systems Programs, "CTM Brief to NAS (U)," presentation to the committee, February 23, 2007, Washington, D.C. (classified); and David W. Lando, Distinguished SLBM Expert, Naval Surface Warfare Center, Dahlgren Division, "CTM: Weapon Effectiveness Presentation (U)," presentation to the committee, July 29, 2007, San Diego, Calif. (classified).

[13]CAPT Terry J. Benedict, USN, Technical Director, U.S. Navy Strategic Systems Programs, "CTM Brief to NAS (U)," presentation to the committee, February 23, 2007, Washington, D.C. (classified); and David W. Lando, Distinguished SLBM Expert, Naval Surface Warfare Center, Dahlgren Division, "CTM: Weapon Effectiveness Presentation (U)," presentation to the committee, July 29, 2007, San Diego, Calif. (classified).

FIGURE 4-15 Damage to a tile roof and the interior ceiling caused by the strike of a meteorite of estimated mass 0.7 kg with an impact speed of less than 500 ft/sec, illustrating the limited damage potential of kinetic energy warheads. Increased speed of impact on structural elements such as walls and roofs does not cause proportional increases in lateral damage, but primarily increases the number of walls and such that can be penetrated. SOURCE: "The Glanerbrug Meteorite Fall," posted by the Dutch Meteor Society, Leiden, The Netherlands, May 18, 1998; see <http://www.xs4all.nl/~dmsweb/meteorites/glanerbrug/glanerbrug.html>.

of targets. The committee recommends that efforts to define the effectiveness of the KEP warhead against targets of interest be continued, as this information will be relevant to several of the envisioned CPGS systems.

The CSM concept envisions the deployment of existing weapons, such as the BLU-108, following deceleration from the reentry conditions. As discussed above, there are technology issues that must be addressed with respect to the dispensing of weapons at high speeds. In addition, the deployed weapons have been well characterized for most target classes of interest, and they are limited in utility for some cases. Some of the difficult issues in addressing mobile targets are described below in the subsection "Mobile Targets."

Fixed Hard Targets

Hardened targets include above-ground hardened structures, shallow underground structures, and deeply buried targets. The issue of lethality against hardened targets was studied recently by the National Research Council.[14] This study concluded that many strategic hard and deeply buried targets can only be attacked directly with nuclear weapons. By their conventional nature, CPGS weapons will not solve the existing shortcomings concerning the attack of these strategic targets.

CPGS systems, with the exception of the initial version of CTM, will be capable of attack of hardened above-ground facilities and shallow underground facilities. The lethality against a hardened target depends on the ability of the weapon to penetrate the hardened features of the target, the ability to fuse the weapon properly, and the ability of the warhead to create the desired effects. Currently, penetrating warheads are limited by structural considerations to impact speeds of approximately 3,000 ft/sec, which sets the upper limit to the capability to penetrate hardened structures. Reducing reentry vehicles speeds to this level also creates technical issues of thermal protection and guidance, navigation, and control, as discussed above.

The more significant challenges lie in the areas of fuse development and effects generation, although these challenges are not significantly different from those for tactical penetrating weapons. Additional R&D is required prior to the deployment of a smart fuse that is reliable and can operate in the presence of easily implemented countermeasures. Techniques to tailor the delivered effects are also required, especially with respect to buried structures containing or manufacturing components and systems for weapons of mass destruction (WMD). The committee recommends robust investigation of techniques necessary to defeat hardened structures, with specific attention paid to the defeat of WMD components and systems.

Mobile Targets

Mobile targets represent one of the most challenging types of target for attack by a CPGS system. The problems are different depending on whether the targets move from time to time but are fixed when weapons arrive (e.g., mobile ICBMs parked somewhere in a dispersal area), or whether they are actually moving when weapons arrive (e.g., a caravan of terrorist leaders moving on a country road). Three technical approaches can be considered for attacking moving targets: (1) terminal sensors, (2) remote sensors and weapon data links, or (3) a combination of both. A terminal sensor can be added to a CPGS system to correct for moderate uncertainties in target location. Incorporating the sensor directly in a high-β RV or hypersonic cruise-missile body presents significant challenges relating to the integration with the thermal protection system. Incorporating the terminal sensor in a deployed weapon, as proposed in the CSM-2 concept, avoids the

[14] National Research Council. 2005. *Effects of Nuclear Earth-Penetrator and Other Weapons,* The National Academies Press, Washington, D.C.

challenging integration with the TPS but requires a high-β RV or cruise missile to slow to an acceptable dispensing speed, which will likely increase its vulnerability in a heavily defended area. The approach of launching a low-β RV to dispense a UAV, as discussed earlier in this chapter, appears an attractive one.

Hitting the *right* moving target is a significant technical challenge. Boost-glide missiles and hypersonic cruise-missile options are assumed capable of deploying BLU-108 submunitions with Skeet warheads. These submunitions are capable of reliably finding motor vehicles (via infrared signatures) and striking them. They cannot discriminate one vehicle from another, however. Originally developed to stop a line of armored tanks, they can be usefully employed if collateral damage is of little importance or if the target vehicle is virtually alone—more than, say, a mile from any other motor vehicle. As a practical matter, in those situations where the target vehicle is moving among other vehicles and collateral damage is to be avoided if possible, human-in-the-loop operations will be required for a CPGS weapon. The committee believes that this will be as true in the year 2020 as it is now. Autonomous target-recognition technology is not advancing at a rate fast enough to solve the significant challenges associated with reliably differentiating one vehicle from another. In large part this is because one cannot predict how, in a given operational situation, the target vehicle will differ from the rest. Perhaps the target vehicle will be identified as "the middle one of three white SUVs traveling in a row." Perhaps it will be identified by its license plate number.

The UAV delivered by the CTM-2 (with UAV) option is assumed to have the necessary capabilities to (1) give remotely located human controllers the visual information that they need to identify the intended target, and (2) permit human control of the UAV, including authorizing it to attack. In this report, it is assumed that CSM-2 and the hypersonic cruise missile could deploy a 2,000 lb UAV capable of the same functions, although it would have less payload and/or range (or loiter time). One possibility for a weapon to arm the UAV is the Hellfire missile. It appears that carriage of at least two Hellfire missiles, at 100 lb each, on either a 3,000 lb or 2,000 lb UAV would be suitable.

Battle Damage Assessment

Near-real-time battle damage assessment is difficult in most strike situations. Better BDA is seemingly always near the top of the commander's list of most-desired capabilities. Unique aspects of CPGS option design, along with the application of modern technology, can potentially give CPGS weapons much better BDA capability than the military is used to with any of its existing strike weapons.

The essence of a CPGS BDA is to carry in the delivery vehicle a deployable device that can hang in the air above the target as the weapon is delivered, take snapshots of the target before and after the target is (one hopes) hit, take a moment to compress the photos into fewer bits of data, and linger in the air long enough to communicate the compressed photos through a UHF satellite link back to the commander who launched the weapon. What makes this concept potentially better than any existing capability is that the camera is very close to the scene, the photo sequence enables a direct comparison of "immediately before" with "immediately after," taking time for photo compression and communication after the strike should enable a high-quality image, and the commander has the photos quickly.

Any CPGS option could, in principle, incorporate this concept, as could any weapon system capable of carrying and releasing the BDA device. For example, in a later version of CTM, one of the reentry bodies could be such a BDA device. In the boost-glide missile or hypersonic cruise-missile concepts, the delivery vehicle itself could possibly perform the function or, probably better, it could dispense a BDA device along with the weapon. The evaluation of this concept is certainly possible outside the context of CPGS, and if it is technically reasonable, incorporating it within a CPGS system would be a valuable addition.

Implications of a Possible Prohibition on Research and Development of Trident-Based Systems

Some in the Congress have called for the Department of Defense (DOD) to proceed in its CPGS program in a manner that excludes systems based on the Trident missile. The intent is to maximize an alleged "bright line" between conventional and nuclear systems. The committee strongly recommends that no such prohibition be adopted or imposed. In particular, it believes that CTM R&D would be very valuable even if the CTM were not deployed. It further believes that the CTM-2's two-stage rocket may sufficiently differentiate it from the full three-stage Trident when tracked by a sophisticated satellite-based system (see Chapter 3 and Appendix H for further discussion of nuclear ambiguity).

Turning the issue around, the committee concludes that if Trident-based R&D were proscribed, the effect would be to delay—perhaps substantially—the development and deployment of any CPGS capability, without actually doing much to reduce the ambiguity problem. It is not that the knowledge gained from CTM (or CTM-2) R&D could not be gained in other ways in time, but rather that much would have to be done from scratch—squandering the many years of base-laying by the Trident Missile Program. In addition, the DOD would have to do substantial component testing for the other options, because the other options simply do not have the advantages of building incrementally from a firm foundation. The least delay (perhaps 2 years or so) would be caused if the alternative system approved were the SLGSM.

TECHNOLOGY READINESS LEVELS AND TIME FRAMES

Many of the advanced technologies that are needed to provide CPGS system capabilities have been under development in laboratory environments or flown as part of the advanced-configuration development programs. An approximate time line for the various CPGS technology options discussed above is shown in Figure 4-16. These technologies build on the development of existing RV and technology development and existing demonstration programs such as the FALCON, X-43, HyFLY, and X-51.

Capabilities beyond the basic ballistic missile and limited boost-glide systems will require the development of more advanced technologies. Boost-glide concepts that use a significant endoatmospheric glide segment will require the operation of a reentry system in a manner that has not been previously demonstrated. One option for development of the CSM involves the exploitation of the reentry vehicle technology developed under the AMaRV program. As envisioned in the initial version of the CSM, the AMaRV vehicles would be scaled up to the size needed to carry the conventional payloads, and the vehicle would be flown with an extended glide segment. The capability of this extended glide segment would be limited by the existing capabilities of thermal protection systems. It is anticipated that this initial capability (labeled CSM-1 in Figure 4-16) would allow for an 800-second glide segment. The committee believes that there are significant technical risks in the operation of existing reentry vehicles in this new flight mode with an extended glide range and that these risks can only be mitigated through system-level demonstration testing.

Very long range glide segments will require the development of new thermal protection systems. The DARPA FALCON flight-test program will explore one approach to the thermal protection system in a reentry vehicle with the high lift-to-drag ratio needed for long-range gliding flight. With demonstrations planned for 2009 and 2010, the FALCON program plans a near-term demonstration of the basic operating characteristics of a vehicle technology that may provide a long-range capability for the CPGS mission. The DARPA FALCON technology is envisioned to feed into a second-generation CSM system (labeled CSM-2 in Figure 4-16). If the FALCON program is successful in demonstrating the technologies necessary for long-range gliding reentry vehicles, the ultimate capability of the CSM system could possibly be developed in a single effort, as opposed to a two-stage process.

The forward-deployed AHW builds on technologies developed under the Sandia Winged Reentry Vehicle (SWERVe) program conducted in the 1980s. For deployment as a CPGS, the SWERVe vehicle technologies will need to be upgraded to enable flight at higher speeds and longer glide range within the atmosphere. The development of the thermal protection system necessary for a forward-deployed boost-glide system will be similar in nature to the development of the TPS for the CSM system. Significant overlap in technology development for boost-glide systems will likely occur if the forward-deployed boost-glide missiles continue to be seriously considered.

The air-breathing Mach 6 missile (hypersonic cruise missile) represents a new class of delivery system, which is immature relative to the ballistic systems.

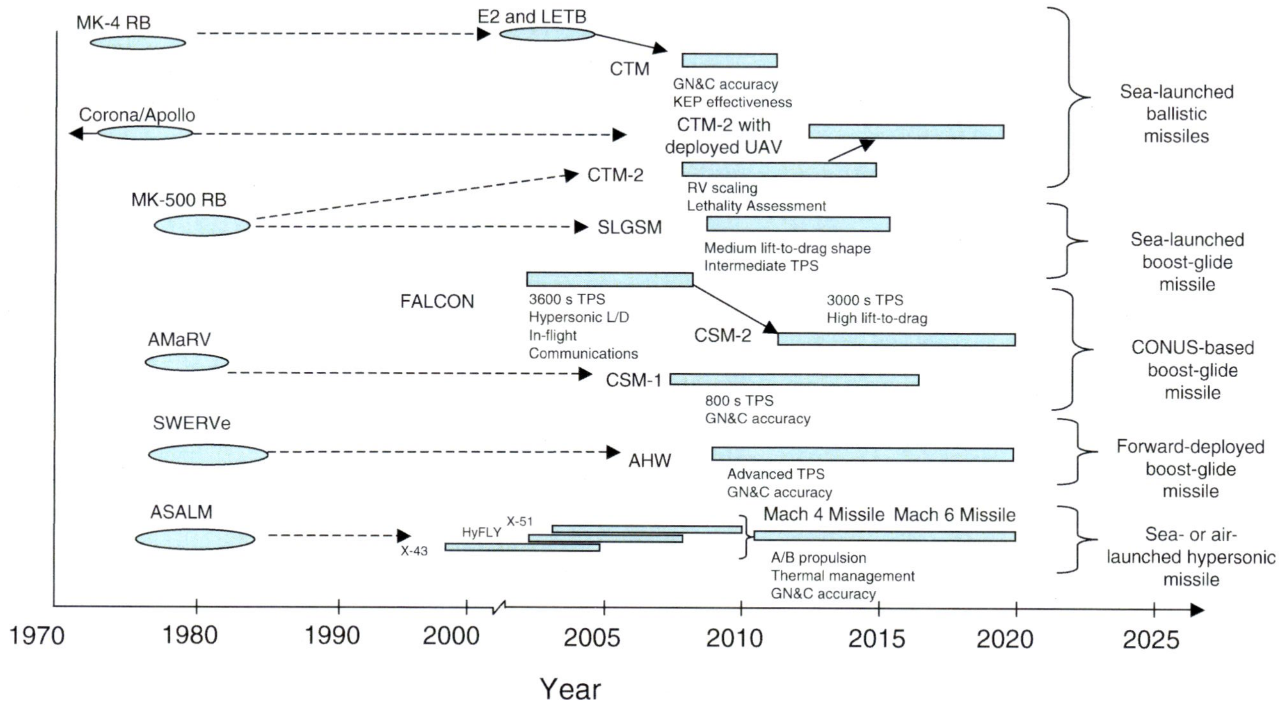

FIGURE 4-16 Technology development time lines illustrating a legacy to all proposed conventional prompt global strike systems. The time lines for future technologies are based on an assumption that investments will continue to be made in critical research and development activities. NOTE: A/B propulsion refers to the four gas generators in the Post Boost Propulsion System on the Trident equipment section (bus) that burn in two stages: the first two burn together called "A" propulsion; once they burn out, a transition is made to the next two gas generators called "B" propulsion. The remaining acronyms are defined in Appendix A.

Technology associated with air-breathing hypersonic propulsion systems has been under development in a laboratory environment for the past 30 years and has recently begun transition to the flight-test environment in programs such as the NASA-funded X-43, DARPA/ONR-funded HyFLY, and USAF/DARPA-funded X-51 programs. These programs have demonstrated or will demonstrate critical aspects of the propulsion technology necessary to enable a Mach 6 cruise missile. Furthermore, the Air Force Research Laboratory is exploring the technologies associated with a Mach 6 hypersonic cruise missile under its Robust Scramjet Technology program.

In 1998, the National Research Council conducted a study evaluating the U.S. Air Force Hypersonic Technology (HyTECH) program.[15] This study concluded that the development of a Mach 6 missile in 2015 was feasible. Although not all recommendations in that report were implemented, the technology readiness of hypersonic cruise missiles is such that this type of capability can be deployed in about 2020.

SUMMARY

The desire for conventional prompt global strike capabilities with the fewest constraints has led to proposals based on ballistic and hypersonic delivery systems. While establishing such capabilities appears feasible, for some concepts it is at the cutting edge of aeronautic technology. The most-effective development of CPGS capabilities will require a spiral technology evolution in which intermediate capabilities are developed and tested, with the results serving as the basis for evaluating and developing more-advanced capabilities. Preliminary testing of the CTM system is an excellent first step in such a development process, because it is strongly connected to proven capabilities and allows key new technologies to be evaluated at relatively low cost. Furthermore, the technologies that must be demonstrated for the success of CTM are common to the success of other proposed ballistic/hypersonic programs for CPGS. The development of CPGS options beyond the relatively limited capabilities of CTM will require additional investments in thermal protection and weapons dispensing. Supporting developmental efforts in these areas as proposed for FALCON and CSM also provides an important step in the future evolution to the most-effective future systems.

FINDINGS AND RECOMMENDATIONS

Finding 1. The command, control, communications, computer, intelligence, surveillance, and reconnaissance (C4ISR) systems needed to enable conventional prompt global strike (CPGS) are only sufficient to meet the CPGS time lines under

[15]National Research Council. 1998. *Review and Evaluation of the Air Force Hypersonic Technology Program,* National Academy Press, Washington, D.C.

limited conditions. Significant additional effort will be required to provide seamless integration of numerous disparate systems and to increase the global coverage of the Digital Point Positional Data Base.

Recommendation 1. The Office of the Secretary of Defense (OSD) should fund the National Geospatial-Intelligence Agency (NGA) to speed production of the Digital Point Positional Data Base (DPPDB) to increase its geographic coverage and to develop a means to determine rapidly and accurately the geographic coordinates of any visually identified point (e.g., from a recent photo taken in the field) when that point lies outside DPPDB coverage.

Finding 2. Conventional Trident Modification (CTM) represents the only near-term option for a CPGS capability, but the system accuracy has not been demonstrated, and the kinetic energy projectile (KEP) warhead will likely be effective against only a subset of candidate targets. In addition, the limited maneuverability of the proposed CTM reentry vehicle will result in an inability to attack targets in many urban and mountainous regions.

Recommendation 2.1. The accuracy of the CTM system needs to be demonstrated in end-to-end system tests. This accuracy demonstration will provide needed information for the CTM system concept, as well as providing important technical information applicable to all CPGS candidate systems.

Recommendation 2.2. Evaluation of the KEP warhead effectiveness should be included in the CTM system tests and defined against candidate target sets. Warheads capable of defeating a wider range of targets should be developed.

Recommendation 2.3. The baseline CTM system concept should include the development and use of reentry vehicles capable of vertical impact to allow the attack of targets in urban or mountainous environments.

Finding 3. A modified version of the Trident missile system, designated in this report as CTM-2, could provide enhanced weapons effectiveness with larger and more flexible payloads.

Recommendation 3. The Navy Strategic Systems Programs Office should conduct a detailed technical assessment of the CTM-2 concept. If deemed feasible, CTM-2 research, development, testing, and evaluation should be provided to address overall system accuracy and weapons effectiveness.

Finding 4. More-advanced concepts for CPGS, such as CTM-2, Conventional Strike Missile (CSM), and Advanced Hypersonic Weapons (AHWs), offer the potential for improved system performance through more flexible payloads and trajectories, but these concepts carry high technical risk that must be mitigated prior to any deployment decision.

Recommendation 4. OSD should fund the technology development of the longer-term CPGS concepts to address the technical issues associated with thermal protection systems; hypersonic aerodynamics and air-breathing propulsion systems; guidance, navigation, and control accuracy; and munitions dispensing systems.

Finding 5. The attack of moving targets and incorporation of battle damage assessment in a CPGS setting will require the development of significant new capability, which could be accomplished with a combination of deployed terminal sensors and weapon data links.

Recommendation 5. OSD should fund the technical evaluation of system concepts to address the attack of moving targets and incorporation of battle damage assessment, including the dispensing of unmanned aerial vehicles from ballistic missiles or boost-glide vehicles.

5

Assessment of Conventional Prompt Global Strike Options—Synthesis

This chapter synthesizes the military issues; the political, international, policy, and doctrinal issues; and the technical assessments presented in Chapters 2, 3, and 4 and presents salient advantages and disadvantages associated with each option. The chapter also discusses some issues that the committee sees in the relationship between conventional prompt global strike (CPGS) and strategic strike in general and how these issues complicate the selection of a longer-term development path at this time.

EVALUATION FACTORS

The chapter begins with a review of several important aspects of the assessment conducted in Chapters 2 through 4—aspects that bear on conclusions to be drawn from this chapter's synthesis of the various issues addressed.

Military Assessment Factors: Requirements and Operational Concepts

The Department of Defense's (DOD's) initial capabilities document (ICD) for CPGS identifies a number of key capabilities that it refers to as "required," such as global reach, promptness, and high probability of achieving the desired effect on the target.[1] These key capabilities were included along with other evaluation fac-

[1] Office of the Secretary of Defense. 2006. *Initial Capabilities Document, Stage for Prompt Global Strike* (U), Washington, D.C., July 29 (classified).

tors in the assessment in Chapter 2 of the potential military effectiveness of each proposed option. The ICD differentiates CPGS from what it calls Next Generation Long-Range Strike by virtue of CPGS's promptness and also by limiting CPGS in volume and persistence of fire. These latter limitations make CPGS (as defined in the ICD) a "niche capability": that is, for prompt fire limited in volume and persistence. The need for CPGS is driven in large part by the goal of a very limited strike in a time of crisis or opportunity, such as to counter terrorist activities. While Conventional Trident Modification (CTM) or CTM-2 is a relatively inexpensive potential means to meet important aspects of the CPGS need and to do so in 3 to 5 years, any longer-term, more versatile option will be a far more expensive national investment that the committee believes, for reasons discussed below, must be put into the broader context of the nation's strategic strike policy and national security strategy. The committee believes that a comprehensive examination of future strategic strike or deep strike would likely result in an allocation of requirements to CPGS weapons different from that defined in the ICD. This committee's skepticism regarding requirements focuses on the use of CPGS weapons in the leading edge of major combat operations rather than on their use in very limited strike in times of crisis or opportunity, as the next paragraphs discuss.

The CPGS joint analysis of alternatives (AoA) currently underway with the U.S. Air Force as lead Service appears to be in competent hands, using good analysis tools and responding faithfully to the CPGS ICD's statement of key capabilities required. It also appears, however, that in adhering to the ICD, the AoA omits any consideration of the possible need for greater firepower. Here, the committee is concerned that, against a large and formidable future adversary, versatile platforms capable of destroying a wide range of targets may be unable to penetrate (or may suffer severe losses in penetrating) the adversary's sophisticated air defenses unless preceded by high-volume defense-suppression attacks. The AoA also takes to heart the ICD's statement that CPGS should be effective against any type of target: The AoA is specifically evaluating each option's capability to destroy a very broad range of targets, including moving targets. While the committee sees this as a laudable goal, it expects that cost and technical risk will militate against requiring the same weapon to have a flight time under an hour *and* to hit moving targets, at least in the next decade or so.

A trade-off study in a broader context encompassing all strategic strike systems that might be considered candidates for development in the next decade or so might reach different conclusions from those reached by the DOD to date. Such a study might result in a concept of operation in which conventionally armed ballistic missiles lethal against a somewhat-limited set of targets (including air defense threats) are first used, followed by long-range, air-breathing vehicles that penetrate into denied-access areas after some defense suppression to provide versatile attack capabilities that reduce dependence on forward-deployed forces.

Political Assessment Factors: Ambiguity and Overflight

Chapter 3 discusses the issue of "nuclear ambiguity," that is, the possible misidentification of a launch of a conventionally armed delivery vehicle as an attack with a nuclear-armed weapon, an issue that has been one of the key issues associated with CPGS weapons. For the reasons explained in Chapter 3, it is the committee's view that there is no "bright line" between conventional and nuclear capabilities based on the physical characteristics of a CPGS option, or on a missile's prior association with an exclusively nuclear role, because any long-range vehicle capable of carrying an effective conventional payload could alternatively carry a nuclear weapon. Accordingly, all CPGS options could have ambiguity issues; those issues cannot be avoided simply by insisting that the system be without a nuclear "legacy." However, it is also the view of the committee that there are many plausible cases in which there would be no risk of the misinterpretation of a CPGS attack (and those cases can be strengthened by various cooperative and unilateral measures), so that the possibility of circumstances in which even a remote probability of misinterpretation would make it imprudent to use a CPGS weapon should not be regarded as a reason to forgo the capability altogether.

Similarly, Chapter 3 assesses the political risk of overflying third-country territory as not being decisive in option selection. Nor, for the reasons stated there, does the committee believe that arms control or other political considerations are decisive reasons not to go forward with CTM (or an alternative CPGS system).

Technical Assessment: Options in Broader Operational Concepts

As noted earlier, in a broader study exploring the full range of strategic strike developments feasible in the next decade or so, requirements might reasonably be apportioned so that a ballistic missile with limited warhead capacity and versatility would handle situations truly requiring flight times of 1 hour or initial penetration into highly defended areas, while long-range air-breathing vehicles with more general utility would handle situations in which a wide range of targets must be struck. This would introduce new options beyond those now being considered for prompt global strike, such as subsonic options that have low technical risk and are capable of more versatile use, making the significant investment more worthwhile.

As discussed in Chapter 4, a large, low-altitude, subsonic cruise missile could be designed with low technical risk for launch from a nuclear-powered guided missile submarine (SSGN) (or a nuclear-powered ballistic missile submarine [SSBN]). It is noteworthy that, for a given number of missiles per launch tube, the subsonic cruise missile can carry a given warhead weight a longer distance than can either an intermediate-range ballistic missile (IRBM) or a hypersonic cruise missile. Its flight time to the desired range, however, will generally be much longer than 1 hour, possibly as long as 8 hours, and it may have difficulty surviving future sophisticated air defenses, although its low-altitude terrain-following capability may give it some advantage over the boost-glide missiles and hypersonic cruise missiles in the latter parts of their trajectories. It should be possible to design the subsonic cruise missile to have considerable warhead versatility.

If the United States were required today to put conventional ordnance on a previously unrecognized target in a corner of the globe distant from current disposition of forces as quickly as possible, the platform most likely to be used for the operation would be a manned bomber (e.g., a B-2) with a great amount of tanker support. Given that part of its long flight time could be in parallel with targeting, mission planning, and decision making, it seems to the committee that this versatile platform type will continue to be considered for the future. For air-breathing weapons delivery systems that must penetrate sophisticated air defenses in the future, the DOD should put greater emphasis on unmanned vehicles, as previous National Research Council (NRC) committees have recommended.[2] A possible combination option is a manned or unmanned bomber that launches either a high- or low-speed cruise missile.

Some basing options for CPGS systems have been considered and rejected. Launching large, long-range weapons from combatant ships would require an expensive and lengthy development (compared with that needed for other basing options), whether it involved a new launcher system designed for multiple legacy ship classes or a new ship class. Manned or unmanned combat aircraft capable of being launched from aircraft carriers would have inadequate range for some CPGS missions.

ASSESSMENT SYNTHESIS

This section summarizes the primary pros and cons of options for a CPGS delivery system, synthesizing the detailed discussion in previous chapters.

Conventional Trident Modification

The Conventional Trident Modification (and possibly its variant CTM-2) is the only potentially viable near-term option available 3 to 5 years from program start. Assuming that tests confirm that CTM would meet planned performance goals, it provides some important capabilities in the relatively near term, is very inexpensive by DOD standards, has low technical risk, and builds on a well-proven and comprehensive system—the Fleet Ballistic Missile force and its infrastructure. Remaining technical uncertainties require that the range of military effectiveness of the CTM be established by testing the delivery accuracy and the effects on target of the munitions prior to deployment. The kinetic energy projectile (KEP)

[2]See, for example, National Research Council, 2005, *Autonomous Vehicles in Support of Naval Operations,* The National Academies Press, Washington, D.C.

warhead (tungsten rods) and the relatively limited payloads that it would carry limit the lethality of CTM to relatively small, relatively soft, and (usually) accurately geolocated targets that are not going to move for at least the "end-to-end" time line for acquisition, decision, and attack. (One possibly significant limitation arising from the flight geometry of the current CTM concept could be eliminated if the Navy's current proposal were modified, possibly without significant program delay, by adding a bent-nose reentry vehicle to enable attacks on the back side of hills or buildings.) While at some point in the future, "peer competitors" may develop missile defenses that would be effective against current U.S. long-range ballistic missiles (at least in small numbers), a Trident-based CPGS system would have an extremely high probability of defense penetration for many years into the future. The committee is confident that the Navy has proposed adequate technical and procedural safeguards against the accidental launch of a nuclear weapon in mixed loads (with the proviso that an expert red team critically review them) and that mixed loads would have no detrimental effect on the nation's nuclear deterrent capability for the next decade or more. Because it is launched from a limited number of SSBN tubes, the CTM is not suitable for high-firepower operations, especially if, as proposed, CTM occupies only two tubes per SSBN. The fall of CTM's early rocket stages may cause overflight issues in certain cases, but the fall is predictable and can be diverted to low-density areas, and the potential damage can be assessed for the decision maker. The CTM is indistinguishable from the nuclear-armed Trident externally or in-flight, but the committee believes that the risks of misinterpretation of a CTM launch as a nuclear attack are very low (even given the stakes) for a wide range of plausible cases and can be lowered still more by cooperative measures and by experience with development, testing, wargaming, exercising—and potentially using—the system. In the committee's view, CTM offers an excellent growth path, as discussed in the next subsection, and its development would provide technology and flight experience valuable for many of the longer-term options should the country choose to pursue them.

Conventional Trident Modification-2

Whether a growth path from CTM or a first-step development itself, the CTM-2 concept proposed in Chapters 2 and 4 would be a significant advance beyond CTM because it would have greater payload capacity. It seems possible that CTM-2 could be available in 5 to 6 years from program start, although a full engineering analysis of the option would be needed to confirm the timescale. The two-stage propulsion of CTM-2 could distinguish it from the three-stage Trident, but with high confidence today probably only when tracked by the systems of the United States itself. Some members of the committee see a possible evolutionary path for CTM-2 that in its next generation could incorporate a reentry vehicle capable of delivering (yet to be developed) air-breathing vehicles into theater for the provision of capability against moving targets and for post-attack assessment and reattack. This evolutionary path appears to those members to be a possible approach to balance technical risk with the rapid fielding of improved CPGS capabilities.

Land-Based Ballistic Missiles

Land-based intercontinental ballistic missiles (ICBMs) have the range, fast time of flight, and survivability for CPGS, as well as relatively low technical risk. One could probably adapt the warhead technology under development for CTM to meet the challenge associated with the land-based ICBM's greater reentry speeds, but this has not been demonstrated. Assuming that this adaptation proved workable, a conventionally armed modification of Minuteman could be operational in perhaps 6 years, given sufficient priority and organizational interest (which does not now exist). Several hundred Minuteman boosters are available in storage. Building more of these boosters for high-volume operations would add substantially to system costs. In the longer term, land-based ICBMs may be able to carry very large conventional warheads capable of penetrating hard and buried (but not deeply buried) targets. Basing conventionally armed ICBMs in Hawaii would reduce overflight issues but would raise "not-in-my-backyard" issues and could be done only with the consent of signatories to (or the termination of) the Strategic Arms Reduction Treaty (START). The ICBM's high-speed terminal approach limits its lethality to a subset of potential target classes, usually accurately geolocated. A conventional Minuteman modification would be indistinguishable from Minuteman externally or in flight, but ambiguity could be resolved cooperatively had there been prior inspection of the vehicle in the silo. The U.S. Air Force decided not to pursue the land-based ICBM option, and the committee supports its decision, as discussed in Appendix I.

The Submarine-Launched Global Strike Missile and Similar Submarine-Launched, Intermediate-Range Ballistic Missiles

The Submarine-Launched Global Strike Missile (SLGSM) or similar submarine-launched IRBM (SLIRBM) has a number of attractive features as a potential mid-term CPGS weapon. Its principal advantage over several other mid-to-long-term options is its relatively low technical risk and relatively early initial operational capability (IOC), perhaps 6 years from program start if the CTM is developed, 8 years if not. The Navy's Strategic Systems Programs (SSP) presented to the committee a point design (specific configuration) for the SLGSM, and Chapter 4 presented the results of a conceptual design study trading range, warhead size, and firepower (number of tubes and number of missiles per tube) applicable to SSGN- or SSBN-launched IRBMs. If based only on SSGNs, the SLGSM point design has a range that can cover most of the areas of interest on the globe. To make it more nearly global, SSP proposes basing it on SSBNs as well. This would restrict the SSBN's operating areas, which could in the future raise issues with respect to the nation's nuclear deterrence capability. The SLGSM's warhead capacity and versatility are less than those of the conceptual CTM-2 and other longer-term options, but it has an advantage over the CTM in these regards. Especially if it adopts terminal countermeasures as defenses improve, the SLGSM should be highly survivable for decades after its introduction. Proposed at two per canister, SLGSM is suitable for high-volume operations, and the committee believes that a three per canister IRBM may be feasible to increase firepower. The SLGSM and other IRBM designs could be differentiated in flight from Tridents, but as with CTM-2, with high confidence today probably only by the systems of the United States itself. In the committee's opinion, nearly constant-heading, largely ballistic trajectories (such as flown by CTM, CTM-2, and land-based ICBMs) have the advantage that they permit a tracker to predict with confidence a weapon's destination, reducing uncertainty about the nature of an attack. The SLGSM's 700-mile glide range increases that uncertainty. A glide range that long also implies about 5 minutes of atmospheric heating, which begins to stress a reentry vehicle's thermal protection system (TPS), as discussed in the next subsection. An SSGN carrying SLGSM could face compromises with its Special Operations Forces mission.

Land-Based Boost-Glide Missiles

The Air Force Space Command has programmed about $480 million over the next 6 years to develop and demonstrate in three flight-tests a hypersonic boost-glide vehicle that it has named the Conventional Strike Missile (CSM). The weapon would be based at Vandenberg and Cape Canaveral Air Force Bases (raising START issues). Another design concept defined in the CPGS AoA is a shorter-range but forward-based (in Guam, Diego Garcia, and Puerto Rico) boost-glide vehicle named Advanced Hypersonic Weapon (AHW). The principal advantages of proposed boost-glide missiles are their (as-yet-unproven) range, warhead capacity, versatility, and maneuverability. These missiles are envisioned to have the capability to deliver KEP or penetrator warheads or to dispense virtually all of the Air Force's air-launched weapons. If these capabilities were proved possible (and the near-real-time targeting necessary in many cases were in fact present), it would make the boost-glide missiles lethal against a very wide range of targets and would often significantly reduce the requirement on target geolocation accuracy. Its ability to alter its trajectory significantly would reduce overflight issues, enable significant cross-range diverting capability for in-flight retargeting, and permit tailoring the end-game approach angle for improved weapons system effectiveness. A boost-glide missile would be easy to differentiate from a ballistic missile late in flight, but its maneuverability raises uncertainty about its destination until then.

The principal disadvantages of the boost-glide missile are its high technical risk, late availability, and high cost. Hypersonic boost-glide vehicle proposals push the performance envelope in the areas of thermal protection and management and the dispensing mechanisms of air-launched weapons. While a ballistic reentry body typically spends about 1 minute in the oxidative hypersonic environment, the CSM under development by the U.S. Air Force would have to fly in it for more than 10 minutes, stressing current technology. In the CSM's next evolutionary step, in order to expand area coverage, it is proposed to fly for nearly 1 hour in that environment, which will require the development of new thermal protection and management technologies. The CPGS AoA is modeling the hypersonic boost-glide vehicles as slowing to Mach 5 to dispense their weapons. This speed seems at the same time aggressively high for dispensing weapons and questionably low for surviving strong area air defenses. A boost-glide missile appears to be an expensive weapon for high-firepower operations.

In the committee's view, the CSM effort planned and funded by the U.S. Air Force is optimistic for a program intended to result in a highly reliable, highly effective presidential-release weapon. The committee's judgment of a prudently scheduled, well-funded program with adequate testing has an IOC of about 2017 for a version of CSM that has a 10-minute-plus glide phase and delivers a penetrator warhead, and (with high technical risk) an IOC of about 2022 for a second version that has nearly a 1-hour glide phase and can dispense a wide variety of air-launched weapons at high speed.

Hypersonic Cruise Missiles

Hypersonic cruise missiles share some of the same promise (e.g., warhead capacity and versatility) and some of the same high technical risks (thermal protection and management and high-speed dispensing of weapons) as those associated with boost-glide missiles. The committee again wonders, as with boost-glide missiles in a weapons-dispensing mode, about the hypersonic cruise missile's survivability against strong regional and local air defenses. The CPGS AoA is modeling a very large air-launched hypersonic cruise missile. A similar missile could also be launched from SSGNs, and Chapter 4 includes a conceptual design study trading range, warhead size, and firepower, complementing the previously mentioned study for IRBMs. Compared to IRBMs, cruise missiles (high- or low-speed) have the advantage that their cross-section can be designed so that multiple missiles fit snugly in an SSGN (or SSBN) tube—for example, three missiles pie-shaped at an angle of 120 degrees to fill the 360-degree circular volume. IRBMs require round rocket-motor chambers, and so multiple IRBMs cannot fill the volume of a tube. This snug fitting enables a hypersonic cruise missile to have about the range of an IRBM when designed for launch from SSGN tubes. As an example, for three missiles per tube and a warhead weight of 1,500 lb, a two-stage IRBM can fly 2,400 miles, including glide range extension, as can a hypersonic cruise missile, including its glide path. Compared with an IRBM or a boost-glide missile, a hypersonic cruise missile can be more easily differentiated from a nuclear-armed ballistic missile in all phases of its trajectory. Of course, it cannot be differentiated from a nuclear-armed cruise missile, and the United

States has a history of such weapons (albeit subsonic), just as it has a history of nuclear-armed ballistic missiles. A disadvantage for the hypersonic cruise missile compared with an IRBM or boost-glide missile is that the hypersonic cruise missile will not fly high enough to escape national sovereignty issues.

CONCLUSIONS

In the near term, CTM (or, with perhaps 2 years' more delay, CTM-2) is the only option. It has the potential to meet important aspects of the immediate need, is inexpensive, and has low technical risk, although its military effectiveness must be evaluated by flight-tests of targeting accuracy and munitions lethality. The committee is confident that issues such as ambiguity, overflight, impact on nuclear deterrence, and accommodation with arms control are manageable and do not constitute a reason to forgo the capability. Moreover, testing and evaluation of the CTM will provide a foundation of information useful for many future options.

In the longer term, submarine-launched ballistic missile options continue to appear attractive, in single or multiple missiles per launch tube, and to offer evolutionary paths that balance technical risk with rapid fielding of improved capabilities. Boost-glide missiles and hypersonic cruise missiles hold promise of versatile capabilities but have high technical risk and questionable survivability against defenses. Land-based ballistic missiles offer little that sea-launched missiles do not, although land-basing avoids concerns about impacting the sea-based nuclear deterrence mission.

In the committee's opinion, the selection of a long-term CPGS delivery system—able to engage more-demanding targets and/or to do so in large volumes—must be put into the broader context of the nation's policy on strategic strike, which requires a new look based on careful analysis and thoughtful debate. In turn, the policy on strategic strike must be consistent with the national security strategy, which a new administration must review beginning in 2009. Any option beyond the CTM or CTM-2 will be an expensive national investment and needs to be well aligned with these yet-to-be-defined national policies. A more comprehensive examination of strategic strike or deep strike and an allocation of requirements to potential future weapons systems would allow a better perspective on CPGS. What CPGS should encompass and which CPGS delivery system option is best are but two of many serious, interrelated questions concerning strategic strike. How much firepower is needed against deep, well-defended targets and how should it be delivered? Can we be confident that air-breathing missiles and boost-glide missiles late in flight will be able to survive future sophisticated air defenses without prior defense suppression? Are the operational advantages of long-range conventionally armed ballistic missiles worth their political risks? What is the future of the SSBN and SSGN? Should they be the basis for a more formidable conventional strategic force in numbers of platforms and in their individual capability?

6

Key Questions and Major Findings and Recommendations

The military, political, and technical implications and issues associated with the research, development, testing, and evaluation (RDT&E) and deployment of U.S. conventional prompt global strike (CPGS) systems, including analysis of leading CPGS options, are discussed in detail in Chapters 1 through 5 of this report. Drawing on these discussions, the committee framed the analysis of CPGS capability by posing the following key questions:

1. Does the United States need CPGS capabilities?
2. What are the alternative CPGS systems, and how effective are they likely to be if proposed capabilities are achieved?
3. What would be the implications of alternative CPGS systems for stability, doctrine, decision making, and operations?
4. What nuclear ambiguity concerns arise from CPGS, and how might they be mitigated?
5. What arms control issues arise with CPGS systems, and how might they be resolved?
6. Should the United States proceed with research, development, testing, and evaluation (RDT&E) of the Conventional Trident Modification (CTM) program[1] and, ultimately, with CTM production and deployment?
7. Should the United States proceed with the development and testing of alternative CPGS systems beyond CTM?

[1]The Conventional Trident Modification program involves the conversion of two Trident II (D5) missiles on each of the U.S. Navy's 12 deployed nuclear-powered ballistic missile submarines from nuclear-armed to conventionally armed.

This concluding chapter comprises a discussion of each of these questions and presents the committee's major findings and recommendations.

1. DOES THE UNITED STATES NEED CPGS CAPABILITIES?

The committee developed a set of credible scenarios and cases with which to assess the feasibility and value of various levels of coverage and promptness and to assess the relative merits of alternative approaches to CPGS. In doing so, it drew on material provided by Department of Defense (DOD) officials, historical experience over the past decade with actual or seriously contemplated strikes, and intelligence projections. The scenarios included, for example, the need to strike a ballistic missile launcher poised to launch a nuclear weapon at the United States or at an ally; an opportunity to strike a gathering of terrorist leaders or a shipment of weapons of mass destruction (WMD) during a brief period of vulnerability; and the need to disable an adversary's command-and-control capability as the leading edge of a broader combat operation.

With the benefit of these scenarios and more specifically defined test cases, the committee concluded that a high-confidence CPGS capability would be valuable; that technical development and assessment should be pursued immediately; and that if system effectiveness is demonstrated, production and deployment should follow as soon as practicable. The committee concluded also that if the DOD's stated goal of achieving "global" strike were to be accepted as a strict criterion, it would rule out potentially attractive options. Long range is an important element of CPGS but not the only factor of interest. Thus, the committee did not interpret the term "global" literally. In contrast, the committee concluded that setting a goal of 1 hour for execution time[2] in a conventional strike was sensible when viewed in terms of feasibility, value, and affordability. But here, too, the goal was not considered as a strict criterion, and some options that would not quite meet the DOD goal were considered in the analysis.

The desire for a CPGS capability has been noted in numerous national defense strategy documents and reports to Congress over the past several years. In February 2007, the Secretary of Defense and Secretary of State submitted to Congress a report that clearly articulated CPGS mission types and made clear the shared DOD and State Department view that CTM is a needed near-term CPGS capability.[3]

At present, U.S. strikes with conventional weapons are conducted primarily through the use of forward-based systems, particularly tactical aircraft and cruise missiles, and with heavy bombers. Effective use of these systems requires that

[2]"Execution time" is the time between the President's order to execute the attack and when the target is affected.

[3]Secretary of Defense and Secretary of State. 2007. "Report to Congress on Conventional Trident Modification (CTM) (U)," Washington, D.C., February 1 (classified).

there be adequate time available to position the aircraft and/or missiles within range of the targets, to conduct detailed mission planning, and, when needed, to provide tanker refueling capability. For distances of about 500 nautical miles (nmi) or more, the flight time alone for current air-breathing vehicles exceeds 1 hour. Accordingly, current forward-based systems can meet the "within 1 hour" criterion for a "prompt" strike only for relatively short distances to targets, and then only if appropriately pre-positioned and with extensive mission-support assets available. The growth of sophisticated air defenses might also present problems for forward-deployed forces, unless attacks by those forces were preceded by effective defense-suppression attacks. Figure 6-1 displays the CPGS "capability gap" that exists at present and indicates that ballistic missiles, either intercontinental or sea-launched, could potentially fill the gap in terms of the time/distance specifications.

Recent U.S. strikes with conventional weapons have included the use of armed unmanned aerial vehicles (UAVs), such as Predator, in attacks against al-Qaeda members in Pakistan. Clearly, if (1) U.S. forces equipped with armed UAVs are deployed sufficiently close to the targets to enable UAVs to reach the targets "promptly" and if (2) local air defenses do not pose an unacceptable threat to the success of the mission, then armed UAVs would provide an option for prompt

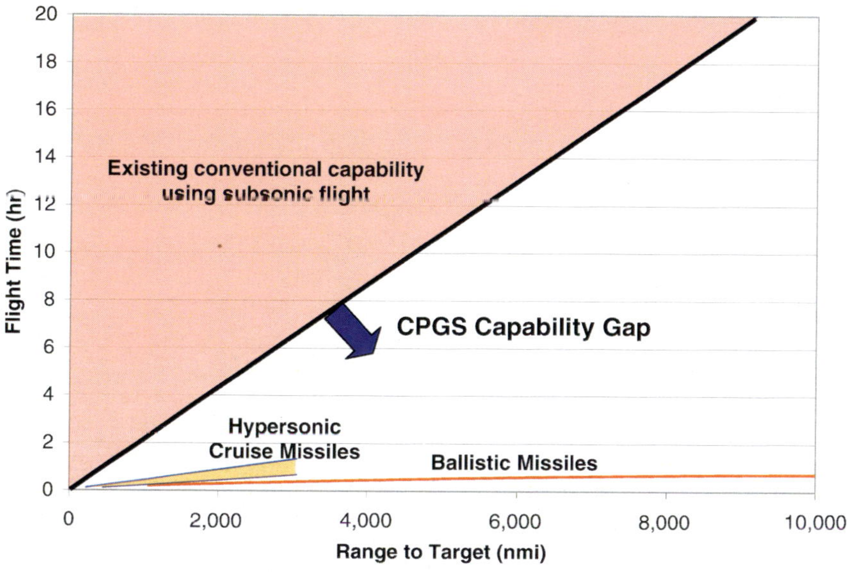

FIGURE 6-1 Conventional prompt global strike (CPGS) capability gap. The time needed to reach a target as a function of range for existing conventional systems contrasted with ballistic missiles and hypersonic cruise missiles.

strike. However, there are many credible scenarios in which these conditions are not met. In addition, CPGS systems employing long-range ballistic missiles with high payloads are projected to be effective against a broader class of targets than could be effectively attacked by armed UAVs.

Major Finding 1. There are credible scenarios in which the United States could gain meaningful political and strategic advantages by being able to strike with conventional weapons important targets that could not be attacked rapidly by currently deployed military assets. In light of the appropriately extreme reluctance to use nuclear weapons, conventional prompt global strike (CPGS) could be of particular value in some important scenarios in that it would eliminate the dilemma of having to choose between responding to a sudden threat either by using nuclear weapons or by not responding at all.

2. WHAT ARE THE ALTERNATIVE CPGS SYSTEMS, AND HOW EFFECTIVE ARE THEY LIKELY TO BE IF PROPOSED CAPABILITIES ARE ACHIEVED?

The U.S. Air Force Space Command serves as the executive agent for the DOD's analysis of alternatives (AoA) for conventional strike and prompt global strike systems. Expected to be completed in May 2008, the DOD AoA examines systems that might be available in the near term, mid-term, and long term: these consist of CTM, with a projected fiscal year (FY) 2010 initial operational capability (IOC)[4] (based on FY 2008 funding); a Conventional Strike Missile (CSM) based in the continental United States (CONUS), with a projected FY 2014/2015 IOC (based on a funded FY 2008 demonstration program); and four conceptual alternatives with projected FY 2020 IOCs. Informed by the DOD's ongoing AoA, the committee elected to analyze seven CPGS system options:[5]

---

[4]"Initial operational capability" is defined in *The Department of Defense Dictionary of Military and Associated Terms* as "the first attainment of the capability to employ effectively a weapon, item of equipment, or system of approved specific characteristics that is manned or operated by an adequately trained, equipped, and supported military unit or force" (Department of Defense, Joint Publication 1-02, April 12, 2001, as amended through October 17, 2007).

[5]Similar to the committee's interim letter report, the final report is based on the committee's collective knowledge as well as on input from other experts, both internal and external to the DOD. Appendix C provides biographical information on the committee members, among whom are technical experts familiar with research, development, and acquisition areas related to strategic strike systems. Accordingly, the committee thought it appropriate in some cases, such as in the case of CTM and CSM, to modify currently proposed DOD CPGS systems in ways that would enhance these systems while at the same time placing their projected capabilities on more realistic paths for potential near-, mid-, and/or long-term utility.

1. *Existing systems.* These comprise tactical aircraft, cruise missiles, other armed unmanned aerial vehicles, and heavy bombers.

2. *Conventional Trident Modification.* CTM is a near-term alternative proposed by the DOD. It involves the conversion of two Trident II (D5) missiles on each of the U.S. Navy's 12 deployed nuclear-powered ballistic missile submarines (SSBNs) from nuclear-armed to conventionally armed. Each converted Trident missile would carry up to four reentry vehicles equipped with advanced navigation, guidance, and control capabilities. Each reentry vehicle would carry a warhead consisting of dispersible kinetic energy projectiles (KEPs).

3. *CTM-2.* CTM-2 is a mid-term alternative conceived by the committee. It would employ conventional warheads on a missile consisting of the first two stages of the three-stage Trident missile. Removal of the third-stage motor would result in increased payload volume and payload options, while still achieving range on the order of 4,000 nmi.

4. *Submarine-Launched Global Strike Missile (SLGSM).* This mid-term to long-term alternative, conceived by the Navy, would build on CTM experience and could carry either multiple KEP warheads or a single, heavier warhead suitable for attack against some hard targets. SLGSMs would be launched from existing nuclear-powered guided missile submarines (SSGNs).

5. *Boost-glide missile (initial version, CSM-1).* CSM-1 is a mid-term alternative and committee modification of the Air Force-proposed Conventional Strike Missile (CSM) concept, which would be launched using a Minotaur ballistic missile (i.e., a modified version of the no longer operationally deployed Peacekeeper). The CSM-1 system would have extended range and substantial maneuvering capability, which would be useful in avoiding undesired overflights of other countries. An initial deployment of CSM-1 would have specified capabilities of 800 seconds of glide time and of delivering either a KEP or a penetrator warhead suitable for use against some hard targets.[6]

6. *Boost-glide missile (second version, CSM-2).* CSM-2 is a long-term alternative and committee modification of the Air Force-proposed CONUS-based missile concept. The specifications for this longer-term, advanced variant of the boost-glide missile would include a 3,000-second glide time to give it a longer range than that of CSM-1, and a capability to slow to speeds appropriate to dispensing multiple munitions and, possibly, to dispensing modules for intelligence,

[6]As discussed in Chapter 4, the time line for the CSM effort planned and funded by the U.S. Air Force is optimistic for a program intended to result in a highly reliable, highly effective presidential-release weapon. In the committee's judgment, a prudently scheduled, well-funded program with adequate testing would have an IOC of about 2017 for an initial version of CSM (which the committee refers to as CSM-1) that has an 800-second glide phase and is capable of delivering either KEP or penetrator warheads, and an IOC of about 2022 for a second version (which the committee refers to as CSM-2) that is proposed to have a 3,000-second glide phase and to be capable of dispensing a wide variety of air-launched weapons at high speed or delivery of KEP or penetrator warheads.

surveillance, and reconnaissance (ISR); battle damage assessment (BDA); and reattack.

7. *Hypersonic cruise missiles.* Hypersonic cruise missiles are long-term alternatives, similar to the AoA's Mach 6 missile, that could be forward-deployed (e.g., on SSGNs or on land) or launched from long-range aircraft. While hypersonic cruise missiles would have only medium range, their specifications would call for considerable capability for terminal-phase dispensing of smart munitions and ISR modules.

Table 6-1 summarizes some of the key attributes of each of these seven alternatives, all of which are analyzed and evaluated in this report. The criteria for evaluation include time to implementation, anticipated cost, delivery accuracy, weapons effectiveness, technical risk, proposed performance in various military scenarios, and contribution to the evolution of long-term CPGS capability.

CTM would be effective against some targets of political and military significance that cannot be struck promptly with existing systems. All alternative CPGS systems, which would to varying degrees overcome the limitations of CTM (e.g., inability to destroy hard or buried targets, limited numbers), would reach IOC later than CTM, as shown in Table 6-1.[7] CTM-2 could be effective against some hard targets (because its payload volume is greater than that of CTM), but the IOC of CTM-2 would lag that of CTM by 1 or 2 years. SLGSM would provide greater payload than CTM and would offer more flexible terminal trajectory together with more firepower (number of tubes and number of missiles per tube) than CTM, but it would be more costly than CTM or CTM-2 and is projected to lag CTM by 3 to 4 years. Boost-glide missiles—such as the CSM-1 and CSM-2—and hypersonic cruise missiles would provide maneuvering capability for larger payloads, but they would have much higher associated costs and developmental uncertainties and would take at least 8 years from now to bring to IOC.

Major Finding 2. Conventional Trident Modification (CTM) has advantages over alternative CPGS systems in its near-term availability, low development cost, low opportunity cost, low technical risk, and minimal required changes in declared policy or doctrine. While CTM has limitations compared with other CPGS alternatives, it would be effective against many targets that current systems could not engage quickly enough, and it is the only CPGS system that could be available in the near term.

[7] A 2005 National Research Council report entitled *Effects of Nuclear Earth-Penetrator and Other Weapons* concluded, among other things, that many strategic hard and deeply buried targets can only be attacked directly with nuclear weapons. See National Research Council, 2005, *Effects of Nuclear Earth-Penetrator and Other Weapons,* The National Academies Press, Washington, D.C.

3. WHAT WOULD BE THE IMPLICATIONS OF ALTERNATIVE CPGS SYSTEMS FOR STABILITY, DOCTRINE, DECISION MAKING, AND OPERATIONS?

The essential policy judgment that must be made in selecting any CPGS system and the doctrine for its use is whether the advantages of the new system would outweigh the disadvantages that it presented. The judgment will depend in part on the type of CPGS system under consideration. Some of the challenges are smaller for CTM than for other CPGS systems because of the CTM program's limited scale, CTM's relatively low cost and technological risk, and its direct lineage from the mature Trident system. However, some of the more advanced CPGS systems (if they prove to be technically feasible) could be designed and deployed so as to reduce ambiguity concerns that arise from the use of delivery vehicles and platforms previously associated with nuclear weapons (see the subsection below).

With the introduction of any significant new military capability, the doctrine for use must be clearly defined by policy makers and clearly understood by military commanders and planners. The use of long-range missiles to deliver conventional weapons promptly and accurately enough to damage meaningful targets requires enabling information (e.g., intelligence, command-and-control, targeting) that is significantly more difficult to obtain than that needed for nuclear weapons or for conventional weapons delivered by manned aircraft.[8] The speed of delivery of enabling information and of a decision to execute must be fast enough to achieve an extremely short overall execution time, yet rigorous command-and-control requirements must be met. The committee believes that a fundamental command-and-control imperative for CPGS systems would be that the weapon could be employed only on the order of the President of the United States. A comprehensive study of the military and diplomatic implications of acquiring and possibly employing CPGS capabilities should precede any deployment and should include the consideration of factors such as the potential for inappropriate, mistaken, or accidental use; the implications for nuclear deterrence and crisis stability (including ambiguity considerations); the impact of overflight and debris; and the implications for arms control and associated agreements.

Major Finding 3. CPGS systems raise policy, doctrine, and operational issues that should be studied comprehensively prior to deployment. The committee's examination of these issues leads it to believe that they could be resolved satisfactorily and that they would not be an obstacle to deployment.

---

[8]Ballistic missile delivery systems were developed for nuclear weapons for which, owing to the large damage area of the weapon, achieving the required accuracy in the placement of the weapon is relatively easy to accomplish. In contrast, for conventional weapons, accuracy of placement (technically referred to in terms of the circular error probable [CEP] or spherical error probable [SEP]) is essential in order to obtain the desired effects on the target.

TABLE 6-1 Summary of Conventional Prompt Global Strike Alternatives Reviewed by the Committee

Alternatives	Origins	Launch Vehicles	Range (Payload-Dependent)[a]	Munitions Payload Capacity[b]	Earliest IOC[c]	20-Year Cost (relative to CTM: billions of 2009 dollars)[d]
Existing systems	USA, USAF, USMC, USN	Cruise missiles, tactical aircraft, and heavy bombers	1,500 to >6,000 nmi	1,000-2,000 lb	Available now	Not applicable
CTM	USN (sea-based)	Trident: D5 (3-stage)	>4,000 nmi	>1,000 lb	2011	1
CTM-2	Committee (sea-based)	Trident: 2-stage	>4,000 nmi[e]	2,000 lb[e]	2013	3
SLGSM	USN (sea-based)	2-stage rocket booster	3,000 nmi	2,000 lb	2014-2015	5-10
Boost-glide missile (CSM-1)	Committee/ USAF (land-based)[f]	Minotaur III	>6,000 nmi	2,000 lb	2016-2020	10-20

Boost-glide missile (CSM-2)	Committee/ USAF (land-based)[f]	Minotaur III	>6,000 nmi (plus additional glide range vs. CSM-1)	2,000 lb	2018-2024	10-25
Hypersonic cruise missiles	USN (sea-based) or USAF (land-based or B-52)	Single-stage rocket booster	2,000-3,000 nmi	1,000-2,000 lb	2020- 2024	10-20

NOTE: Acronyms are defined in Appendix A.

[a]Data on range and payload for CTM, SLGSM, CSM-1, CSM-2, and hypersonic cruise missile options are extracted from Amy F. Woolf, 2007, *Conventional Warheads for Long-Range Ballistic Missiles: Background and Issues for Congress,* CRS Report to Congress, Congressional Research Service, Washington, D.C., June 19, pp. 10-12, 24-26.

[b]The reader is cautioned that direct mass-to-mass comparisons of munitions capacity do not reflect weapons effectiveness. Different types of munitions will have different weapon impact for the same mass.

[c]The reported initial operational capability (IOC) data in this table are the committee's best estimates based on information presented to the committee and the experience of committee members, assuming an authorization date of 2008. Actual IOCs for all but the CTM are likely to be later for many reasons, including delays in decision making, the time required to stand up program offices, and unanticipated problems in systems engineering.

[d]The 20-year cost estimates are based on contractor briefings. The numbers quoted are imprecise estimates of costs relative to the projected cost for CTM.

[e]The committee-generated CTM-2 concept would have a larger payload capability due to the throw weight and volume freed up by removing the third-stage motor of a Trident missile. Range, however, would be somewhat lower depending on payload.

[f]CSM-1 and CSM-2 are committee modifications of the Air Force-proposed CSM and CONUS missile concepts, respectively.

Areas of comprehensive study should include the potential for inappropriate, mistaken, or accidental use; the implications for nuclear deterrence and crisis stability (including "ambiguity" considerations); the impact of overflight and debris; and the implications for arms control and associated agreements.

4. WHAT NUCLEAR AMBIGUITY CONCERNS ARISE FROM CPGS, AND HOW MIGHT THEY BE MITIGATED?

Nuclear ambiguity is the most frequently raised objection to proceeding with CTM.[9] This concern is often described as arising only in the case of CPGS systems that (like CTM) would use delivery vehicles and platforms previously associated with nuclear weapons. However, the concern applies to varying degrees to any CPGS system, since any vehicle capable of delivering a conventional weapon to long range and with high speed and accuracy would also be effective in delivering a nuclear weapon.

If another country, for example Russia or China, were to detect the launch of one or more conventionally armed long-range missiles from a deployed SSBN, how might it interpret the event?

There are two aspects, logically and practically distinct, of the nuclear ambiguity issue. The first is the possible misinterpretation by an observing nation of a conventional strike on a third party as a nuclear strike on its own territory. The second is the possible misinterpretation by an observing nation of one or more conventionally armed missiles headed toward its territory as a nuclear attack. The ambiguity issue is more significant in the second case. The committee's analysis of the nuclear ambiguity issue focused on the following questions:

- Who would be able to detect the launch?
- If a foreign nation were to detect the launch, would it be able to identify correctly the missile type and estimate the trajectory of the missile or the reentry vehicles?
- If a launch were detected by a foreign nation, what would happen?
- Would that nation's nuclear forces and surveillance systems be alerted, and if so what would be the consequences?
- Would a "retaliatory" strike be ordered?
- Even if there were no immediate adverse effects, what long-term reactions might be triggered?

---

[9]In a letter dated February 16, 2007, to Dr. Ralph J. Cicerone, President of the National Academy of Sciences, Senators Daniel K. Inouye and Ted Stevens, Chairman and Ranking Member, respectively, of the Senate Committee on Appropriations, Subcommittee on Defense, stated that "there was widespread, but not universal, agreement [in the Senate] that the Congress should not proceed with the conventional Trident program [and that] critical to the opposition was a belief that the Trident option proposed the most difficult challenge of ambiguity." This letter is reproduced in Appendix B.

These questions are stated simply, but the answers do not lend themselves to such simplicity. Nevertheless, the committee's judgment is as follows:

- In the next 5 years or so, only Russia will be able to detect the launch of a ballistic missile from a sea-based platform. China might obtain such capability soon thereafter.
- If a foreign nation were able to observe the launch of a ballistic missile, it might also be able to determine the missile type. If it were to track the early flight of the missile, it would be able to predict the subsequent ballistic trajectory of the reentry vehicles. Predicting the course of a maneuverable system would be far more difficult.
- The reaction of a nation observing the launch would depend on the context. For example, is the observing nation an adversary of the United States or an ally? Does the missile appear to be headed to a target on its territory? Is this event occurring in the midst of a period of conflict or in a time of relative peace?
- Command and surveillance systems would likely be fully alerted and, depending on the context, military forces (possibly including nuclear forces) might also be brought to higher alert status.
- The committee believes that the risk of the observing nation's launching a nuclear retaliatory attack is very low. A foreign nation would be extremely unlikely to believe that the United States was starting a nuclear war with only a handful of missiles, and that nation would have every incentive, in its own interest, to determine definitively the character of the attack before responding.
- This risk could be reduced even further by means of cooperative measures, such as providing information to bilateral partners about the CPGS system, its operation, and the doctrine for its use; immediately notifying of launches against countries; and installing devices (such as continuous monitoring systems) to increase the confidence that conventional warheads had not been replaced by nuclear warheads.[10]
- The possibility of conditions in which misinterpretations would be plausible is not, in the committee's judgment, a valid reason to forgo the CPGS capability for those many other cases in which the risk of misinterpretation is negligible.
- Substantive (as opposed to rhetorical) international reactions to the U.S. acquisition and possible use of CPGS capabilities probably would include countermeasures intended to protect valuable potential targets and might include increased emphasis on acquiring comparable conventional strike capabilities. The intensity of any such reactions would be expected to depend to a substantial degree on the scale of the U.S. CPGS deployment or use. For example, the detection of the launch of a large number of conventionally armed missiles might be

[10]Appendix H provides a discussion of some cooperative efforts (both technical and nontechnical) that can be applied toward mitigating nuclear ambiguity.

interpreted as an attempted "disarming first strike" against a nation's strategic forces. The committee believes that the CTM program, as currently envisioned, is sufficiently small in scale to make it unlikely that international reactions would be of strategic significance.

Major Finding 4. Nuclear ambiguity is an understandable concern regarding CTM and, to varying degrees, all other CPGS systems. Nuclear ambiguity cannot be eliminated simply by avoiding a "legacy" nuclear system, such as Trident. The risk of a CPGS attack being misinterpreted and leading to a nuclear attack on the United States could be mitigated and managed through readily available mechanisms. The benefits of possessing a limited CPGS capability, such as that provided by CTM, outweigh the risks associated with nuclear ambiguity.

5. WHAT ARMS CONTROL ISSUES ARISE WITH CPGS SYSTEMS, AND HOW MIGHT THEY BE RESOLVED?

Both the Intermediate-Range Nuclear Forces (INF) Treaty of 1987 and the Strategic Arms Reduction Treaty (START) of 1991 contain provisions that would apply to certain CPGS systems. The Moscow Treaty of 2002 (Strategic Offensive Reductions Treaty, or SORT) requires reduction and limitation of strategic nuclear warheads, but it does not constrain non-nuclear warheads and, therefore, does not affect any CPGS system.

The INF Treaty had been signed in 1987 by the United States and the former Soviet Union; after the breakup of the Soviet Union, Russia, Belarus, Kazakhstan, and Ukraine assumed the obligations of the former Soviet Union. That treaty prohibits flight-testing, production, and deployment of ground-launched ballistic missiles and ground-launched cruise missiles with ranges between 500 and 5,500 kilometers, regardless of warhead type. It places no restrictions on manned aircraft, air-launched or sea-launched systems, or on ground-launched systems with ranges less than 500 or greater than 5,500 kilometers. The INF Treaty is of unlimited duration.

START requires the United States and Russia to limit their deployed strategic arsenals to no more than 6,000 warheads, with no more than 4,900 on intercontinental ballistic missiles (ICBMs) and submarine-launched ballistic missiles (SLBMs), in accordance with agreed counting rules. START covers all ICBMs and SLBMs and their associated launchers, including new types of ballistic missiles, and no distinction is made between nuclear-armed and non-nuclear-armed ballistic missiles. START is scheduled to remain in force until December 5, 2009, but even if not extended or renegotiated, it is unlikely that its restrictions will disappear entirely.

With these arms control issues in mind, an obvious question for some readers may be why some of the Air Force land-based Minuteman III missiles should not be converted from nuclear-armed to conventionally armed in a manner similar to the Navy CTM program. After thorough investigation of the Minuteman option, the committee concluded that, although technically viable, it is not a realistic contender (for reasons described in Appendix I). In short, the required renegotiation of START, combined with "not-in-my-backyard" issues, presents substantial challenges to deploying ICBMs in locations such as Hawaii or Guam, particularly in the near term. These challenges are judged to be significantly more formidable than the obstacles to CTM deployment. Neither minimizing the number of conventionally armed ICBMs to be deployed in Hawaii or Guam nor using mobile units rather than silos altered the committee's conclusion.[11]

Major Finding 5. Neither CTM nor CTM-2 is in conflict with the Intermediate-Range Nuclear Forces (INF) Treaty of 1987 or the Strategic Arms Reduction Treaty (START) of 1991 because (1) the INF Treaty applies only to ground-launched missiles, and (2) even though the START rules for counting compliance with its limits on missiles, launchers, and warheads would apply to CTM and CTM-2, total planned U.S. deployments of systems subject to those limits are sufficiently below START limits to allow for the envisaged deployments of CTM or CTM-2. Other CPGS systems could raise arms control issues if the INF Treaty and START remain in force or are renegotiated without including provisions that would permit the deployment of these systems.

6. SHOULD THE UNITED STATES PROCEED WITH RESEARCH, DEVELOPMENT, TESTING, AND EVALUATION OF THE CTM PROGRAM AND, ULTIMATELY, WITH CTM PRODUCTION AND DEPLOYMENT?

As indicated earlier, the committee believes that there exist plausible scenarios in which CTM would be a valuable and reasonably early addition to U.S. military capabilities, and that CTM should be deployed as quickly as possible, provided that the system's planned effectiveness is demonstrated in tests and that international and political concerns are appropriately mitigated and managed. Accordingly, the committee believes that the United States should proceed with CTM RDT&E to demonstrate system effectiveness (especially with respect to achieving accuracy on the order of meters and lethal effects on the classes of targets of interest), while concurrently pursuing cooperative measures by which to

[11]In the committee's judgment, any proposed land-based, conventional ICBM system, such as one based on the no longer operationally deployed Peacekeeper, would face the same near-term concerns for deployment as those outlined in Appendix I for Minuteman. These near-term concerns include the likely renegotiation of START, thereby raising the possibility of complications, delays, and uncertainties, as well as the political, economic, and procedural difficulties of introducing new and secure operational bases for strategic missiles.

address and mitigate the concerns of Russia and other countries. The committee believes also that the proposed RDT&E program for CTM should be expanded to include the consideration of reentry vehicles capable of diving vertically (in order to attack targets located, for example, on the far side of hills or buildings). Full-scale production and deployment of CTM should proceed only after tests have confirmed that the projected effectiveness of the system has been achieved.

As for the potential for meeting challenges facing CTM, the committee believes (1) that deployment of CTM (i.e., a total of 24 Trident II [D5] missiles converted from nuclear-armed to conventionally armed) would not adversely affect the nation's nuclear deterrent for the foreseeable future; (2) that the Navy has proposed reasonable technical and procedural safeguards against accidental or mistaken launch of a nuclear-armed Trident missile from an SSBN loaded with both nuclear-armed and conventionally armed missiles (and that these safeguards should be strengthened even further in accordance with lessons learned from the Air Force incident in 2007 involving the mistaken transport of nuclear-armed cruise missiles on a B-52 bomber,[12] and that these safeguards should be tested to ensure their effectiveness); (3) that because the fall of early rocket stages from the CTM would be predictable, any issues associated with overflight could be adequately taken into account by decision makers; (4) that CTM would survive and penetrate likely missile defenses for the foreseeable future; and (5) that the nuclear ambiguity problem inherent in CPGS systems can be adequately managed in the case of CTM.

Finally, the committee notes that CTM, in comparison with alternative CPGS systems under consideration by the DOD and Congress, would be available significantly sooner, with far less cost and far less development risk and uncertainty.

Major Finding 6. The military and political issues associated with CTM appear to be manageable. The CTM research, development, testing, and evaluation (RDT&E) program would be useful not only in demonstrating the effectiveness of CTM as a precondition to full-scale production and deployment, but also in providing technology and flight experience for more advanced CPGS systems. Failure to proceed with CTM RDT&E could significantly delay progress on other CPGS systems.

7. SHOULD THE UNITED STATES PROCEED WITH THE DEVELOPMENT AND TESTING OF ALTERNATIVE CPGS SYSTEMS BEYOND CTM?

CPGS options employing SLBMs are attractive in the near term (CTM), mid-term (CTM-2), and long term (SLGSM) and offer evolutionary paths that balance

[12]Joby Warrick and Walter Pincus. 2007. "Missteps in the Bunker," *Washington Post,* September 23.

technical risk with the rapid fielding of improved capabilities.[13] In the long term, boost-glide missiles and hypersonic cruise missiles offer versatile capabilities, but they have high cost and high technical risk and may be less able than ballistic missiles are to penetrate defenses. Land-based ballistic missile options offer little that SLBMs do not, but their higher payloads would enable them to carry conventional warheads better suited to attacking some hard targets, albeit with larger technical risk in obtaining the necessary targeting accuracy.

The committee believes that the United States should proceed with technology development to support the mid- and long-term CPGS options; however, any decisions to proceed with full-scale testing or beyond should be made in the broader context of the nation's policy on strategic strike. The policy on strategic strike must be consistent with the U.S. national security strategy, which the next presidential administration will review in 2009. Development and deployment of any CPGS option beyond CTM or CTM-2 would require a very large national investment.

Major Finding 7. CTM-2 is worth exploring as a mid-term successor to CTM. Of the long-term alternative CPGS systems, the Submarine-Launched Global Strike Missile (SLGSM) appears to have the lowest technical risk and to offer important capabilities, such as the ability to launch from existing, dedicated, conventional strike platforms (nuclear-powered guided missile submarines, or SSGNs). Boost-glide missiles and hypersonic cruise missiles have higher technical risk but, if demonstrated, could provide some advantages beyond submarine-launched ballistic missiles (SLBMs) (e.g., in payload, versatility, and maneuverability).

MAJOR RECOMMENDATIONS

Based on its analysis of CPGS, the committee offers the following recommendations.

[13]Whether a growth path from CTM or a first-step development itself, the CTM-2 concept would be a significant advance beyond CTM because it would have greater payload capacity. It seems possible that CTM-2 could be available in 5 to 6 years from program start, although a full engineering analysis of the option would be needed to confirm the timescale. CTM-2's two-stage propulsion could distinguish it from the three-stage Trident, but with high confidence today probably only when tracked by the United States' own systems. Some members of the committee felt strongly that if a possible evolutionary path for CTM-2 were taken, the next generation (beyond CTM-2) could incorporate a reentry vehicle that could deliver (yet-to-be-developed) air-breathing vehicles into theater so as to provide capabilities against moving targets and for post-attack assessment and reattack. Such an evolutionary path is an important approach toward balancing technical risk with rapid fielding of improved CPGS capabilities.

Near Term (1 to 2 Years)

In the near term (the next 1 to 2 years), the committee recommends that the following be done:

- Fund CTM research, development, testing, and evaluation (RDT&E) at a level sufficient to achieve early deployment if tests confirm system effectiveness. In parallel, develop doctrine and policies for use of the system, including ensuring the availability of the necessary intelligence and other enablers, and for decision-making procedures.
- Fund exploration of the potential of CTM-2 and, if it is deemed promising, fund RDT&E for CTM-2.
- Fund technology development to explore longer-term CPGS weapon delivery options.
- Initiate programs to improve targeting, planning, and decision making to support CPGS capability, and establish decision points about where to go beyond CTM.
- Initiate what will likely be a multiyear study of concepts and doctrine for potential larger-scale CPGS deployments, recognizing that going beyond the niche capability anticipated currently would raise profound issues that have not yet been adequately explored and debated.

Mid-Term (3 to 5 Years)

The committee recommends that for the mid-term (the next 3 to 5 years), the following be carried out:

- If CTM is demonstrated to be effective, particularly its ability to achieve the necessary accuracy and warhead performance, fund full-scale production and deployment of CTM.
- If CTM-2 is demonstrated to be significantly more effective than CTM, fund the production and deployment of CTM-2 as a follow-on to CTM.
- Continue to fund technology development for the more promising longer-term CPGS options.
- Work with allies, Russia, and others to mitigate policy and international concerns associated with a U.S. CPGS capability, including establishing cooperative measures to reduce ambiguity risks.
- Conduct a comprehensive examination of strategic strike (nuclear and conventional)—across a full spectrum of scenarios—after review of the national security strategy by the next administration, and determine the implications for the development and possible deployment of advanced CPGS systems and associated enabling systems.

Long Term (Beyond 5 Years)

The committee recommends that in the long term (beyond 5 years), the following be done:

- Explore a range of CPGS options and fund RDT&E of the most promising options at levels sufficient to improve weapon effectiveness (especially maneuverability, accuracy, range, and lethality).
- Continue to work closely with U.S. military and diplomatic representatives, coalition forces, allies, and others to mitigate policy and international concerns as they may arise.

Appendixes

A

Acronyms and Abbreviations

AHW	Advanced Hypersonic Weapon
ALCM	air-launched cruise missile
AMaRV	advanced maneuvering reentry vehicle
AoA	analysis of alternatives
AOR	area of responsibility
ASALM	advanced strategic air-launched missile
ASAT	antisatellite
AT&L	Acquisition, Technology and Logistics
BDA	battle damage assessment
BGV	boost-glide vehicle
C2	command and control
C3	command, control, and communications
C4ISR	command, control, communications, computers, intelligence, surveillance, and reconnaissance
CALCM	conventional air-launched cruise missile
C/C	carbon-carbon
CEP	circular error probable
CET	commander evaluation test
CM	cruise missile
CMM	Conventional Minuteman Modification
COCOM	combatant commander
CONOPS	concept of operations
CONUS	continental United States

CPGS	conventional prompt global strike
CSAR	combat search and rescue
CSM	Conventional Strike Missile
CTM	Conventional Trident Modification
DARPA	Defense Advanced Research Projects Agency
DASO	demonstration and shakedown (missile)
DOD	Department of Defense
DPPDB	Digital Point Positional Data Base
E2	Enhanced Effectiveness
F2T2EA	Find, Fix, Track, Target, Engage, and Assess
FALCON	Force Application and Launch from CONUS
FBM	Fleet Ballistic Missile
FOBS	fractional orbit bombardment system
FOC	final operational capability
FY	fiscal year
GBI	ground-based interceptor
GN&C	guidance, navigation, and control
GPS	Global Positioning System
HDBT	hard and deeply buried target
HE	high explosive
HTV	hypersonic test vehicle
HyFLY	Hypersonic Flight Demonstration
HyTECH	Hypersonic Technology
ICBM	intercontinental ballistic missile
ICD	initial capabilities document
IMINT	imagery intelligence
IMU	inertial measurement unit
INF	Intermediate-Range Nuclear Forces
INS	Inertial Navigation System
IOC	initial operational capability
IRBM	intermediate-range ballistic missile
ISR	intelligence, surveillance, and reconnaissance
JDAM	Joint Direct Attack Munition
KEP	kinetic energy projectile

L/D	lift-to-drag
LETB	Life Extension Test Bed
MAC	multiple all-up-round canister
MaRB	Maneuverable Reentry Body
MaRV	Maneuverable Reentry Vehicle
MCO	major combat operations
MLRB	medium lift reentry body
MM	Minuteman
MTCR	Missile Technology Control Regime
NA	not applicable
NASA	National Aeronautics and Space Administration
NCA	National Command Authority
NGA	National Geospatial-Intelligence Agency
NR	not required
NRC	National Research Council
NRO	National Reconnaissance Office
NSB	Naval Studies Board
ONR	Office of Naval Research
OSD	Office of the Secretary of Defense
PG	precision-guided
PGM	precision-guided missile
PGS	prompt global strike
PK	Peacekeeper
QDR	Quadrennial Defense Review
RB	reentry body
R&D	research and development
RDT&E	research, development, testing, and evaluation
ROE	rules of engagement
ROK	Republic of Korea
ROM	rough order of magnitude
RV	reentry vehicle
SEP	spherical error probable
SIOP	Single Integrated Operational Plan
SLBM	submarine-launched ballistic missile
SLCM	sea-launched cruise missile
SLGSM	Submarine-Launched Global Strike Missile

SLIRBM	submarine-launched, intermediate-range ballistic missile
SM-3	Standard Missile-3
SOF	Special Operations Forces
SORT	Strategic Offensive Reductions Treaty
SSBN	nuclear-powered ballistic missile submarine
SSGN	nuclear-powered guided missile submarine
SSN	nuclear-powered attack submarine
SSP	Strategic Systems Programs
START	Strategic Arms Reduction Treaty
STRATCOM	U.S. Strategic Command
SWERVe	Sandia Winged Reentry Vehicle
TEL	Transport Erector Launcher
TLE	target location error
TPS	thermal protection system
TRL	technology readiness level
TTPs	tactics, techniques, and procedures
UAV	unmanned aerial vehicle
U/G Facil.	underground facility
UHF	ultrahigh-frequency
Unk	unknown or unspecified
USA	United States Army
USAF	United States Air Force
USMC	United States Marine Corps
USN	United States Navy
WMD	weapons of mass destruction

B

Letter from Senators Inouye and Stevens

United States Senate
COMMITTEE ON APPROPRIATIONS
WASHINGTON, DC 20510–6025
http://appropriations.senate.gov

February 16, 2007

Dr. Ralph J. Cicerone
President
National Academy of Sciences
500 Fifth Street
Room NAS 215
Washington DC 20001

Dear Dr. Cicerone:

We appreciate that the Academy of Sciences is proceeding to examine the conventional Trident missile in response to the funding and guidance included in the FY 2007 Defense Appropriations Act. In its deliberation on the bill, the Senate eliminated the budget request for research and development of the conventional Trident program. While proponents of the program sought to restore funding for the program, they were defeated on a vote of 67-31. We would like to share our views on the reasons why the Senate took this position, and the ultimate conference outcome.

The Administration recommended funding to establish a conventional Trident program to afford the President an option to attack targets quickly without resorting to nuclear weapons. We understand the Trident option was selected because the Administration placed a self imposed requirement to deploy this capability within 24 months. While we agree finding alternatives to resorting to nuclear weapons is a worthy goal, the Congress has not specifically endorsed the mission of prompt global strike – and certainly not with a timeline of 24 months.

We were advised by several current and former military leaders who questioned whether prompt global strike was a valid concept given the actual state of our targeting capability and intelligence resources. Moreover some bristled at the notion that the Trident submarine would be the preferred platform for a prompt global strike if the mission were approved.

As we discussed this with our Senate colleagues it became apparent that there was widespread, but not universal, agreement that the Congress should not proceed with the conventional Trident program. Some of our colleagues opposed any program, while others believed that a land based alternative, or perhaps an air based option, would be preferable to the Trident.

Critical to the opposition was a belief that the Trident option proposed the most difficult challenges of ambiguity. The proponents argued that a Trident missile could be easily retrofitted to provide a very quick strike option. However, the fact that one would not be able to differentiate between a conventional missile launch and a nuclear missile launch from a Trident submarine was viewed with particular concern by those of us who opposed the program.

Moreover, while the Defense Department had reviewed the matter internally and had directed reviews by others, we believed the Congress needed a truly independent study to examine the manifold issues involved in the decision to proceed with the prompt global strike mission and in particular the conventional Trident missile option.

We recommended the National Academy of Sciences conduct such a study because of our belief that the Academy could provide a review completely independent of the views and desires of the Defense Department.

The conference agreement on the FY 2007 Defense Appropriations bill ratified most of the recommendations of the Senate. The conferees agreed that questions remain about the conventional Trident missile and that an independent study would be advisable. In a change from the Senate position, the conferees agreed that $20 million should be made available to continue to study – and develop – technologies which might be used in a prompt global strike mission, to the extent that the Congress ultimately decides to approve such a plan. However, the conferees insisted that the funds should not be used to develop projects which were only for the Trident option.

We hope the information that we have provided here is helpful to the Academy in framing the issue. We look forward to working with you as you proceed. Please let us know if there is additional information we could provide to you on this matter.

Sincerely,

Daniel K. Inouye
Chairman
Committee on Appropriations
Subcommittee on Defense

Ted Stevens
Ranking Member
Committee on Appropriations
Subcommittee on Defense

C

Committee and Staff Biographies

Albert Carnesale, *Chair,* is chancellor emeritus and professor at the University of California at Los Angeles (UCLA). He served as chancellor from July 1, 1997, to June 30, 2006. As chief executive officer, he led an institution of more than 38,000 students and 27,000 faculty and staff; was responsible for all aspects of the university's mission of education, research, and service; managed an enterprise with an annual budget of $3.5 billion; and served as principal spokesperson for the university community. Dr. Carnesale holds professorial appointments in UCLA's School of Public Affairs and Henry Samueli School of Engineering and Applied Science. His research currently focuses on issues in international affairs and security and in higher education. Prior to assuming the chancellorship of UCLA in 1997, Dr. Carnesale was at Harvard University for 23 years, serving as provost of the university from 1994 to 1997. He held the Lucius N. Littauer Professorship of Public Policy and Administration at Harvard's John F. Kennedy School of Government, where he served as academic dean (1981-1991) and dean (1991-1995). His earlier career included positions in the private sector and in government. Dr. Carnesale has represented the U.S. government in high-level negotiations on defense and energy issues (including the Strategic Arms Limitation Talks, SALT I), and has consulted regularly for several government agencies and companies. He holds bachelor's and master's degrees in mechanical engineering and a Ph.D. in nuclear engineering, has been awarded three honorary doctorate degrees, and is a fellow of the American Academy of Arts and Sciences and a member of the Council on Foreign Relations.

Paul J. Bracken is professor of management and political science at the Yale University School of Management, where his research interests include strategy, technology, and war, as well as business, government, and globalization. Prior to joining Yale, Professor Bracken was a member of the senior staff at the Hudson Institute for 10 years, where he directed the management consulting arm of the institute. He has also taught in corporate education programs internationally, for Gazprom, Lukoil, and other energy companies dealing with a new risk map; for the Israeli Defense Force; and for many U.S. corporations. Dr. Bracken has authored several books, including *Fire in the East: The Rise of Asian Military Power and the Second Nuclear Age* (1999). He is a member of the Council on Foreign Relations and currently serves on the Chief of Naval Operations Executive Panel.

Paul K. Davis is a principal researcher in the RAND Corporation and a professor of policy analysis in the Pardee RAND Graduate School. He previously served a 5-year tour as RAND's corporate research manager for defense and technology planning. His recent monographs discuss capabilities-based planning, strategic planning for counterterrorism, the implications of modern decision science for the support of policy makers, and the use of portfolio-management methods for the analysis of acquisition options. Prior to joining RAND, Dr. Davis was a senior executive in the Office of the Assistant Secretary of Defense for Program Analysis and Evaluation. Before that, he had worked on strategic arms control (SALT II) at the Arms Control and Disarmament Agency and on strategic technology at the Institute for Defense Analyses. Dr. Davis has served on numerous committees of the National Research Council (NRC) and the Defense Science Board. He is a former member of the NRC's Naval Studies Board.

Steve Fetter is dean of the School of Public Policy at the University of Maryland, where he has been a professor since 1988. He has been a visiting fellow of the Center for International Security and Cooperation at Stanford University, the Center for Science and International Affairs at Harvard University, the Plasma Fusion Center at the Massachusetts Institute of Technology, and the Lawrence Livermore National Laboratory. He has also served as a special assistant to the Assistant Secretary of Defense for International Security Policy and was a Council on Foreign Relations Fellow in the State Department. Dr. Fetter's research interests span a wide range of fields, including nuclear weapons, arms control, and nonproliferation policy; nuclear power and the health effects of radiation; and climate change and energy supply. Dr. Fetter has served on numerous scientific boards and advisory committees, including the Department of Energy's Nuclear Energy Research Advisory Committee, the National Academy of Sciences (NAS)/ National Academies' Committee on International Security and Arms Control, and the NRC Committee on the Effects of Nuclear Earth-Penetrator and Other Weapons.

John S. Foster, Jr. (NAE), is an independent consultant whose current clients include Northrop Grumman Space Technology, Sikorsky Aircraft Corporation, NineSigma, and Defense Group, Inc. Dr. Foster retired from TRW, Inc., as vice president for science and technology in 1988 and continued to serve on the board of directors of TRW until 1994. From 1965 to 1973, he served as director of defense research and engineering for the Department of Defense. Prior to this, he served as director of the Lawrence Livermore National Laboratory and associate director of the Lawrence Berkeley National Laboratory. Dr. Foster began his career at the Radio Research Laboratory of Harvard University and spent 2 years as an adviser to the 15th Air Force on radar and radar countermeasures in the Mediterranean theater of operations. A member of the National Academy of Engineering (NAE), Dr. Foster has served on numerous scientific boards and advisory committees, including the NRC Committee on Department of Defense Basic Research.

Eugene Fox, Major General, USA, retired in 1989 after 33 years of service in the U.S. Army. His last active-duty position was as deputy director of the Strategic Defense Initiative Office. During his Army career, General Fox commanded field artillery and air defense units from platoon to brigade level, instructed in a Service school, and served in various capacities in the acquisition of Department of Defense weapons systems, including several years as program manager. Since his military retirement, General Fox has served as a defense consultant for various companies and government agencies. Although he has no prior National Research Council experience, he did serve as a reviewer on the 2001 NRC report entitled *Naval Forces' Capability for Theater Missile Defense*.

Alec D. Gallimore is professor of aerospace engineering and associate dean of the Horace H. Rackham Graduate School at the University of Michigan, where his research interests include electric propulsion, plasma diagnostics, space plasma simulation, electrode physics, and hypersonic aerodynamics and plasma interaction. He has extensive design and testing experience with a number of electric propulsion devices, including Hall thrusters, ion engines, arcjets, 100-kW-class steady magnetoplasmadynamic thrusters, and multimegawatt pulsed coaxial plasma accelerators. Professor Gallimore has implemented a variety of probe, microwave, and optical/laser plasma diagnostics, and he has graduated 15 Ph.D. students and 11 M.S. students in the field of electric propulsion. He has written more than 170 archival journal articles and conference papers and 2 book chapters on electric propulsion. He is also the director of the NASA-funded Michigan Space Grant Consortium and is a member of the U.S. Air Force Scientific Advisory Board. Professor Gallimore has served on numerous scientific boards and advisory committees, including the NRC Committee on Future Air Force Needs for Survivability.

Richard L. Garwin (NAS, NAE, IOM) is an emeritus fellow at the IBM Thomas J. Watson Research Center. A member of the NAS, NAE, and Institute of Medicine (IOM), his expertise is in experimental and computational physics and includes contributions to nuclear weapons design, instruments and electronics for nuclear and low-temperature physics, computer elements and systems, superconducting devices, communications systems, the behavior of solid helium, and the detection of gravitational radiation. Dr. Garwin has served on numerous scientific boards and advisory committees, including the President's Science Advisory Committee (1962-1965); the 1998 Rumsfeld Commission—the Commission to Assess the Ballistic Missile Threat to the United States; and the NRC Committee on the Effects of Nuclear Earth-Penetrator and Other Weapons. In addition, he has been an active member of the NAS/National Academies' Committee on International Security and Arms Control since 1980. He currently consults for the Los Alamos National Laboratory and Sandia National Laboratories and is an active member of the JASONs. He has written extensively on nuclear-weapons-related issues over the course of several decades, particularly on the question of maintaining the nuclear stockpile under a comprehensive test ban regime. Until August 2001, he chaired the Department of State's Arms Control and Nonproliferation Advisory Board. He is a fellow of the American Physical Society and the American Academy of Arts and Sciences and a member of the American Philosophical Society.

Eugene E. Habiger, General, USAF, retired in 1998 as commander-in-chief, United States Strategic Command, where he was responsible for all U.S. Air Force and U.S. Navy strategic nuclear forces supporting the national security strategy of strategic deterrence. In this position, he established military-to-military relationships with his Russian counterparts, fostering confidence-building and openness. After his retirement, General Habiger was appointed the U.S. Department of Energy's director of security and emergency operations. He also served as a distinguished fellow and policy adviser with the University of Georgia Center for International Trade and Security, where he assisted with the center's international programs aimed at preventing weapons proliferation and reducing nuclear dangers. He has served as the president and chief executive officer of the San Antonio Water System and has served on the Nuclear Threat Initiative's board of directors.

David V. Kalbaugh is an independent consultant, having retired in 2005 as assistant director for programs at the Johns Hopkins University Applied Physics Laboratory (JHU/APL), where he was responsible for the oversight and coordination of all laboratory technical programs. During his career at JHU/APL, Dr. Kalbaugh was head of the Power Projection Systems Department, where he was responsible for programs in strike warfare, defense communications, and information operations. His background is in tactical missile and precision strike systems. He joined JHU/APL in 1969 and was involved in the development of the

Tomahawk cruise-missile system at its inception. In addition to his supervisory and management duties, Dr. Kalbaugh has taught for more than a decade in JHU's Whiting School of Engineering. He has served on numerous scientific boards and advisory committees, including participation in tasks for the Under Secretary of Defense for Acquisition and for the Program Executive Officer for Theater Air Defense. In addition, Dr. Kalbaugh served as co-chair of the NRC Committee on C4ISR for Future Naval Strike Groups. He is a former member of the NRC's Naval Studies Board.

L. David Montague (NAE), an independent consultant, is retired president of the Missile Systems Division at Lockheed Martin Missiles and Space. A member of the NAE, he has a background in military weapons systems, particularly in regard to the guidance and control of submarine-launched weaponry. His experience has focused on both tactical and strategic strike systems, as well as on the requirements for, development of, and policy issues related to defense systems to protect against weapons of mass destruction. Mr. Montague is a fellow of the American Institute of Aeronautics and Astronautics and has served on numerous scientific boards and advisory committees, including task forces for both the U.S. Army and and the Defense Science Board, and the NRC Committee on Distributed Remote Sensing for Naval Undersea Warfare. He is a current member of the NRC's Naval Studies Board.

Robert B. Oakley is Distinguished Visiting Fellow at the Institute for National Strategic Studies at the National Defense University (NDU). Prior to joining NDU, he served in various assignments in the U.S. Foreign Service, beginning in 1958; he first served in Khartoum, Sudan, and then in the Office of the United Nations Political Affairs in the Department of State. His subsequent posts included Abidjan, Ivory Coast; Saigon, Vietnam; Paris, France; the U.S. Mission to the United Nations; and Beirut, Lebanon. Ambassador Oakley became deputy to the Assistant Secretary of State for East Asia and Pacific Affairs and was then posted to Zaire, and later to Somalia, as the U.S. Ambassador. He was appointed director of the State Department Office of Combating Terrorism. Prior to joining the National Security Council staff as Assistant to the President for the Middle East and South Asia, Ambassador Oakley became a fellow at the Carnegie Endowment for International Peace. After his retirement as Ambassador to Pakistan in September 1991, he became associated with the U.S. Institute of Peace as a coordinator of the Special Program in the Middle East Peacekeeping and Conflict Resolution. Ambassador Oakley has served on numerous scientific boards and advisory committees, including the NRC Committee for an Assessment of Non-Lethal Weapons Science and Technology.

Walter B. Slocombe is a partner at the law firm of Caplin & Drysdale, where he specializes in civil litigation, defense trade, and exempt organizations. He has served in numerous government positions throughout his career, including as under secretary of defense for policy (1994 to 2001) and director of the DOD Task Force on Strategic Arms Limitation Talks (SALT) (1977-1981). He has also served as senior adviser for national defense in the Coalition Provisional Authority for Iraq. In 2004, Mr. Slocombe was appointed by the President of the United States to the Commission on the Intelligence Capabilities of the United States Regarding Weapons of Mass Destruction. Mr. Slocombe serves on numerous scientific boards and advisory committees, including the International Advisory Board of the Geneva Centre for the Democratic Control of Armed Forces and the NRC Committee on the Policy Consequences and Legal/Ethical Implications of Offensive Information Warfare. He is also a former member of the Strategic Air Command Technical Advisory Committee.

William D. Smith, Admiral, USN, retired in 1993 after 38 years of service in the U.S. Navy. Now a senior fellow at both the National Defense University and the Center for Naval Analyses, Admiral Smith has a background in Navy planning, programming, budgeting and in operational issues principally within the submarine force. His last assignment was as U.S. military representative to the North Atlantic Treaty Organization Military Committee in Brussels, Belgium. In addition, he served in a number of high-ranking capacities for the Chief of Naval Operations, such as Deputy Chief of Naval Operations for Logistics and Navy Program Planning (1987-1991) and as director, Fiscal Management Division/Comptroller of the Navy (1985-1987). His decorations include the Defense Distinguished Service Medal, the Distinguished Service Medal with Gold Star, the Legion of Merit with Three Gold Stars, the Meritorious Service Medal with Gold Star, and the Navy Commendation Medal. Admiral Smith has served on numerous scientific boards and advisory committees, including as chair of the Naval Space Panel Review for the Under Secretary of the Navy and co-chair of the NRC Committee on the Navy's Needs in Space for Providing Future Capabilities. He is a former member of the NRC's Naval Studies Board.

John P. Stenbit (NAE) is an independent consultant whose expertise includes system architectures for complex military and communications systems and systems engineering of information systems. Mr. Stenbit formerly served as Assistant Secretary of Defense for Networks and Information Integration and as chief information officer of the DOD. Prior to serving in the DOD, he served as executive vice president at TRW, Inc. A member of the NAE, Mr. Stenbit has served on numerous scientific boards and advisory committees, including as a member of the NRC Committee on C4ISR for Future Naval Strike Groups and as a current member of the NRC Committee on the "1,000-Ship Navy"—A Distributed and Global Maritime Network. He is a current member of the NRC's Naval Studies Board.

David M. Van Wie is director of the Precision Engagement Transformation Center at the Johns Hopkins University Applied Physics Laboratory, where his principal research interests are in the field of aerospace vehicle design and development, with emphasis on propulsion systems and advanced aerodynamics for supersonic and hypersonic flight vehicles. Dr. Van Wie also holds appointments as research professor in the Department of Mechanical Engineering at the Johns Hopkins University and as lecturer in the Department of Aerospace Engineering at the University of Maryland. He received B.S., M.S., and Ph.D. degrees in aerospace engineering from the University of Maryland and an M.S. in electrical engineering with emphasis on radar and communications systems from the Johns Hopkins University. Dr. Van Wie has has served on numerous scientific boards and advisory committees, including the NRC Committee on Future Air Force Needs for Survivability.

Robert H. Wertheim (NAE), Rear Admiral, USN, retired in 1980 after 35 years of service in the U.S. Navy. During his naval career, Admiral Wertheim served as director of Navy Strategic Systems Programs, where he was responsible for the research, development, production, and operational support of the Navy's submarine-launched ballistic-missile systems—Polaris, Poseidon, and Trident. After retiring from the Navy, he spent 7 years as senior vice president of science and engineering at Lockheed Corporation, and since then has been an independent consultant. A member of the NAE, Admiral Wertheim has served on numerous scientific boards and advisory committees, including as a member of the NAS/National Academies' Committee on International Security and Arms Control (1989-1997) and the NRC Committee on the Effects of Nuclear Earth-Penetrator and Other Weapons. He has also served as a member of the Joint Department of Defense/Department of Energy Advisory Committee on Nuclear Weapons Surety and on the National Security Panel of the University of California President's Council.

Ellen D. Williams (NAS) is a professor in the Department of Physics at the University of Maryland, as well as director of the Materials Research Science and Engineering Center at the university. A member of the NAS, she is also a fellow of the American Academy of Arts and Sciences, and in 2001 she was the recipient of the American Physical Society's David Adler Lectureship Award. Dr. Williams serves on numerous scientific boards and advisory committees, including the National Security Panel of the University of California President's Council, the editorial board of *Nano Letters* of the American Chemical Society, and the NRC Committee on Nanotechnology for the Intelligence Community. She also currently serves as a member of the NRC Board on Army Science and Technology.

Staff

Charles F. Draper is director of the NRC's Naval Studies Board. Before joining the NRC in 1997, he was the lead mechanical engineer at S.T. Research Corporation, where he provided technical and program management support for satellite Earth station and small satellite design. He received his Ph.D. in mechanical engineering from Vanderbilt University in 1995; his doctoral research was conducted at the Naval Research Laboratory (NRL), where he used an atomic-force microscope to measure the nanomechanical properties of thin-film materials. In parallel with his graduate student duties, Dr. Draper was a mechanical engineer with Geo-Centers, Inc., working on-site at NRL on the development of an underwater x-ray backscattering tomography system used for the nondestructive evaluation of U.S. Navy sonar domes on surface ships.

D

Statement of Task and Congressional Language

The conference report accompanying the FY 2007 Department of Defense Appropriations Act (H.R 5631/Public Law 109-289) directed the National Academy of Sciences to conduct a study to analyze the mission requirement for using existing Trident II (D5) missiles with conventional payloads to provide a prompt global strike capability and, where appropriate, consider and recommend alternatives that meet the prompt global strike mission in the near term (1-2 years), mid-term (3-5 years), and the long term.[1] The study should include analyses of the military, political, and international issues associated with each alternative. The study should consider technology options for achieving desired objectives as well as mitigating policy concerns.

This 15-month study will produce two reports: (1) a letter report following the second full committee meeting that summarizes the requirements and supporting enablers for a conventional prompt global strike capability and recommends a near-term option or options to provide this capability; and (2) a comprehensive report that addresses the full terms of reference as outlined above.

Following is the congressional language regarding prompt global strike contained in the conference report requesting this study:

> The budget request includes $127,000,000 to demonstrate the feasibility of using existing TRIDENT II (D-5) missiles with conventional payloads to provide a prompt global strike capability. The conferees believe that fundamental issues about the requirement for and use of this weapon must be addressed prior to

[1] *Making Appropriations for the Department of Defense for the Fiscal Year Ending September 30, 2007, and for Other Purposes: Conference Report to Accompany H.R. 5631,* H. Rept. 109-676, pp. 227-228, 109th Cong., 2d sess. (September 25, 2006).

determining the efficacy of this program. Therefore, the conferees are providing $5,000,000 in Research, Development, Test and Evaluation, Defense-Wide for a study to be conducted by the National Academy of Sciences to analyze the mission requirement and, where appropriate, consider and recommend alternatives that meet the prompt global strike mission in the near term (1-2 years), the mid-term (3-5 years), and the long term. The study should include analyses of the military, political and international issues associated with each alternative. The study should consider technology options for achieving desired objectives as well as mitigating policy concerns. The study is due to the congressional defense committees by March 15, 2007. In addition, the conferees are providing $20,000,000 in Research, Development, Test and Evaluation, Navy for developmental efforts under the Conventional Trident Modification program. These funds should be used to focus on those developmental items which are common to all the global strike alternatives until the completion of the study and a determination has been made on the best course of action in this matter.[2]

[2]*Making Appropriations for the Department of Defense for the Fiscal Year Ending September 30, 2007, and for Other Purposes: Conference Report to Accompany H.R. 5631,* H. Rept. 109-676, pp. 227-228, 109th Cong., 2d sess. (September 25, 2006).

E

Interim Letter Report to Congress

THE NATIONAL ACADEMIES
Advisers to the Nation on Science, Engineering, and Medicine

Naval Studies Board
500 Fifth Street, NW
Washington, DC 20001
Phone: 202 334 3523
Fax: 202 334 3695
E-mail: nsb@nas.edu
www.nationalacademies.org/nsb

May 11, 2007

The Honorable Daniel K. Inouye, Chairman
The Honorable Ted Stevens, Ranking Member
Subcommittee on Defense
Senate Committee on Appropriations
Dirksen Senate Office Building, SD-119
Washington, DC 20510

The Honorable John P. Murtha, Chairman
The Honorable C. W. Bill Young, Ranking Member
Subcommittee on Defense
House Committee on Appropriations
Capitol Building, H-149
Washington, DC 20515-6018

Dear Senators Inouye and Stevens and Representatives Murtha and Young:

The conference report accompanying the 2007 Department of Defense (DOD) Appropriations Act (H.R. 5631/Public Law 109-289) requested that the National Academy of Sciences conduct a study to analyze the mission requirement for using existing Trident II (D5) missiles with conventional (i.e., non-nuclear) payloads to provide a prompt global strike capability and, where appropriate, consider and recommend alternatives that meet the prompt global strike mission in the near-, mid-, and long-term. The study requested analyses of the military, political, and international issues associated with each alternative and asked that the committee consider technology options for achieving desired objectives as well as mitigating policy concerns.[1]

Accordingly, the National Research Council (NRC) established the Committee on Conventional Prompt Global Strike Capability in February 2007.[2] The committee's statement of task charges it to produce two reports over a 15-month period: (1) a letter report, following the second full committee meeting, that summarizes the requirements and supporting enablers for a conventional prompt global strike capability and recommends a near-term option or options to provide this capability; and (2) a comprehensive report that addresses the issues as outlined above. This letter constitutes the committee's first report.[3]

The findings and recommendations in this interim letter report are based on the committee's collective knowledge as well as input from other experts, both internal and external to the DOD. The committee has received a number of very helpful briefings as well as information in other forms[4] that it believes constitute a sufficient basis for its initial findings and recommendations. As explained below, however, this initial information received has served to raise important questions concerning which the committee is not yet prepared to offer definitive views.

Accordingly, this first report is very much an interim letter report that neither addresses in its entirety any one element of the statement of task nor reaches final conclusions, but rather touches on aspects of immediate considerations associated with Fiscal Year 2008 (FY 08) funding. The purpose of this interim letter report is to provide advice to Congress that can be used during the FY 08 appropriations and authorization process as the committee continues to investigate the military, technical, political, and international issues associated with the Conventional Trident Modification (CTM) program and potential alternatives for providing a conventional prompt global strike (CPGS) capability. The committee will continue its work during the coming months and expects to finish drafting a final report by early 2008. The final report will address in detail all of the elements in the study's statement of task.

BACKGROUND ON CONVENTIONAL PROMPT GLOBAL STRIKE

As discussed below, there are a variety of circumstances in which it could serve U.S. national objectives to be able to strike targets very rap-

[1]The statement of task is provided in enclosure A.

[2]The committee roster is provided in enclosure B.

[3]Information about the independent review of the committee's report under the supervision of the NRC's Report Review Committee is provided in enclosure C.

[4]A summary of the data-gathering sessions to date is provided in enclosure D.

idly, with high accuracy and high confidence of reaching the target, and with necessary military effect, but without using nuclear weapons. Modern technology, in particular the Global Positioning System (GPS), makes it possible, in principle, to achieve high probabilities of success with a far more limited number of conventional weapons than in the past. In many circumstances, forward-deployed assets—such as tactical aircraft, cruise missiles, long-range bombers, and unmanned aerial vehicles—make it possible to strike targets with very high accuracy and in sufficiently short times (particularly taking into account the other factors that lengthen the timeline between detection of a target and weapon impact—including evaluation of intelligence, decision to attack, confirmation of geolocation, and input into guidance systems—many of which can occur concurrent to readying or prepositioning of a weapon system).

Taking the long view, however, it is clear that the United States cannot always rely on having forward-deployed forces in the right place at the right time. The question then becomes how timely conventional strikes must be in order to be effective. The time between a strike's launch and its impact on the target is, of course, only one of the many factors in the overall time needed. These factors—not all of which can be run in parallel—include intelligence collection, analysis, and dissemination; discussion of options by the appropriate decisionmakers; transmission and receipt of orders; precise geolocation of targets and transfer of this information to the weapons systems; and detailed mission planning and preparation of weapons systems for launch. A comprehensive effort to make speedier response possible should be a part of any effort to achieve CPGS. However, there is no doubt that the time from launch to impact on a target is also a factor, and the DOD has concluded—and the committee concurs—that situations might arise for which achieving promptness in that variable (launch to effective strike accomplished within an hour or so of an execution order) would add meaningfully to the nation's military capabilities. Among currently available delivery systems, only long-range ballistic missiles can reach targets in very remote areas with very high speed and little or no vulnerability to defense—and to date, long-range ballistic missiles have only been equipped with nuclear warheads.

As discussed in the 2006 Quadrennial Defense Review (QDR), the DOD has assessed potential conventional prompt global strike options, including sea- and land-based ballistic missiles and advanced technologies such as hypersonic glide vehicles. The QDR called for deployment, within 2 years, of an "initial capability to deliver precision-guided conventional warheads using long-range Trident Submarine-Launched Ballistic Missiles [SLBMs]."[5] The associated CTM program calls for the conversion of two

[5]Department of Defense. 2006. 2006 *Quadrennial Defense Review,* Washington, D.C., February 6.

Trident II (D5) missiles on each of the 12 deployed strategic ballistic missile submarines to non-nuclear warheads for conventional prompt global strike. Each converted Trident missile is expected to carry up to four warheads.[6]

Congress, however, has raised several concerns about CTM, specifically uncertainties as to whether (1) a CTM launch could be misinterpreted as the launch of a nuclear weapon, (2) possession of the capability might lead to unwarranted strikes by the United States, and (3) intelligence support is or would be sufficient to support the effective use of CTM. It has also raised technical questions regarding the characteristics of a conventional ballistic missile warhead and its effects against a range of targets, including ones that are mobile, hardened, or deeply buried. As a result, the FY 07 defense bill funded only a small portion of the President's budget request for CTM and limited use of the funding to efforts that are not unique to CTM as such but also support other options for CPGS.

KEY FINDINGS

Purpose of and Need for CPGS

The committee was asked to summarize in its letter report the requirements for CPGS. The Secretary of Defense and the Combatant Commander, U.S. Strategic Command (USSTRATCOM), have indicated clearly their belief in the requirement for CPGS (within an hour or so from launch), as soon as the United States can have it. Also, a report on CPGS recently submitted to Congress by the Secretary of Defense and the Secretary of State[7] and reviewed by the committee quite clearly articulates mission types that both Defense and State believe justify CTM as a needed near-term capability. That is, they agree that a valid requirement exists. The committee shares the view that CTM, if demonstrated to be effective, would be a valuable addition to U.S. military capabilities.

Most broadly, the CTM program should be seen as part of an evolutionary process in which the United States would develop long-range non-nuclear weapons with launch-to-impact times that previously were possible only for nuclear weapons. There are numerous potential missions for CPGS, but in the view of the committee, they can be separated into two distinct categories, as outlined below in the first finding.

---

[6]CAPT Terry Benedict, USN, Technical Director, Navy Strategic Systems Programs, presentation to the committee, March 22, 2007.

[7]Secretary of Defense and Secretary of State. 2007. "Report to Congress on Conventional Trident Modification (CTM) (U)," February 1 (classified).

Finding 1: There appear to be at least two, clearly distinguishable purposes for a CPGS capability:

a. **Very limited (e.g., use of one to four weapons), time-critical strike in a time of crisis or opportunity, such as to counter terrorist or rogue state activities or to attack a terrorist leader; and**
b. **A strike at distant, time-critical targets as the leading edge of major combat operations.**

These two purposes are quite different in their operational requirements and also in how the political environment affects the decision to use specific types of weapon delivery, especially ballistic missiles. Moreover, the supporting enablers for missions associated with these two purposes are quite different and put very different demands on command, control, communications, computers, intelligence, surveillance, and reconnaissance (C4ISR) support.

Based on the information presented to the committee thus far, it appears that the supporting enablers for CPGS can be summarized as follows: intelligence support, mission planning, target development, and decisionmaking. Because of the depth and breadth of this topic, as well as the perceived levels of classification associated with it, a comprehensive analysis of the supporting enablers for CPGS, including prior and post intelligence of a strike, is not attempted here but will be addressed in the committee's final report.

For the reasons discussed below, it appears that CTM is most applicable for the very limited time-critical strike. The United States currently has no truly global non-nuclear capability for that purpose, promptness might be essential, and very few weapons are involved.

CPGS for Very Limited Strike in Time of Crisis or Opportunity

The committee was provided several briefings and references that cited either directly or inferentially the *Washington Post* editorial by former Defense Secretaries James Schlesinger and Harold Brown[8] describing a compelling scenario in which the need for CPGS is evident: the United States has learned of a terrorist group's plan to transport a nuclear weapon, and the opportunity to intercept the shipment is both urgent and fleeting. In this scenario, there are no U.S. military forces close to the expected shipping point and one weapon type in the U.S. arsenal can reach the point in time—a nuclear-armed ballistic missile. Clearly, the nation would benefit from hav-

[8]Harold Brown and James Schlesinger. 2006. "A Missile Strike Option We Need," *Washington Post,* May 22.

ing a conventional option in this case. Another scenario involves a rogue state preparing to launch a ballistic missile with a nuclear warhead from a location that current conventional forces could not reach with sufficient speed. And yet another oft-cited and plausible scenario is one in which the United States has learned that a top terrorist leader will be at a certain place at a given time and again the nation has no conventional forces capable of striking that place at the right time.

As Congress has noted, the C4ISR architecture must be capable of supporting a CPGS weapon. In each of the above very limited strike scenarios, it is possible that detailed attack and targeting preparations will have been made—such as georegistration (determining the latitude, longitude, and elevation of the target), planning for minimization of collateral damage, assessment of the vulnerability of the target to the warhead-type, and the triple checking of intelligence. In these cases, it is plausible that it would be important to be able to strike very quickly so that—while decisionmaking and preparations would always take some time—when the triggering event or opportunity arose, execution of the strike would be as rapid as technology could support. Moreover, there might be instances, even with such targets, where pre-planning could be such that flight time would become the critical element in the ability to respond quickly enough.

It is also quite possible—perhaps more probable—that preparations would not have been made ahead of time, in which case the need for rapid georegistration would be as great as the need for rapid weapon delivery. Achieving sufficiently accurate and reliable georegistration within minutes is a daunting challenge. Similarly, rapid decisionmaking (presumably by the President) would require expedited abilities to assess the risk of collateral damage and other risks peculiar to the use of a CPGS weapon. The committee has not yet had adequate opportunity to understand fully the DOD's or the White House's capabilities in this regard.

Given the pace of terrorism's spread and the consequent uncertainty about where terrorist operations will occur, coupled with the proliferation of weapons of mass destruction, a truly global capability may soon be required, if it is not required today.

Inventory requirements for a CPGS weapon would depend on the range of the weapon, the number of warheads needed to accomplish a militarily effective strike, and the concomitant basing plan required to achieve prompt global coverage. The committee suspects that very limited strikes using a CPGS weapon in, say, the first decade after its fielding would likely number at most a few dozen. For example, only a few terrorist leaders would merit use of such a weapon. In any case, the committee concluded that the very limited strike mission would require a smaller weapon inventory, and also calls for an earlier initial operational capability, than would be needed for uses leading into major combat operations (MCO).

The Ambiguity Issue

The possibility that a very limited strike in a time of crisis or opportunity could be mistaken as a nuclear attack, especially with use of a ballistic missile for strike delivery, must be soberly assessed as decisions are made with respect to both fielding a weapon and using it. While ambiguity issues may be mitigated by cooperative measures, any CPGS option, including CTM, should be designed in both hardware and operational terms to minimize the possibility of misinterpreting intent, specifically taking into consideration detection and tracking capabilities anticipated in the world over the next 10 to 15 years. Although the ambiguity problem may not be as significant as some believe, the committee thinks that it merits serious consideration. Indeed, the ambiguity between nuclear and conventional payloads can never be totally resolved, in that any of the means for delivery of a conventional warhead could be used to deliver a nuclear warhead. It remains to be seen whether nuclear-related security or cooperative measures might ease the problem.

Continuing study and analysis of tradeoffs will be necessary during CTM development. As an example of an issue meriting further study, it is possible that a different concept of deployment could improve strike effectiveness and further reduce ambiguities. The committee has not yet had the opportunity to engage DOD on these issues, to examine the operational tradeoffs in any detail, or to consider fully the opportunities and pitfalls of consultations and negotiations with the Russians and others to eliminate the ambiguity problem, but it plans to do so in the coming months.

The political environment determining what may be at stake can vary for a very limited strike. It is the committee's understanding that the administration has discussed with other nations the ambiguity concerns and is addressing the ambiguity problem in other aspects of its consideration of the CPGS issue. In its further work, the committee will address this issue in detail. Since the immediate issue is what further development work is appropriate in the coming year, the ambiguity issue does not have to be finally resolved at this time.

CPGS for Leading Edge of Major Combat Operations

The leading edge of more substantial operations (i.e., MCO) is the second, distinguishable purpose for a CPGS capability. DOD representatives presented scenarios where the use of CPGS weapons in the first hours of what might become major combat operations to attack distant target sets could be of great operational importance. In some cases, it might be necessary to reach far inland very quickly to cripple an adversary's essential warfighting capabilities before they could be used with potentially decisive effect against U.S. or allied forces. Also, new systems would have an advantage over airbreathing platforms and weapons (e.g., tactical aircraft, cruise missiles, long-range bombers, and unmanned aerial vehicles) for reaching even less-time-urgent distant targets in that they would not be vulnerable to the sophisticated air defenses that might be faced in the future. New systems would also not require the same defense suppression, tanking, and other support, which increases the total force's exposure and vulnerability, limits tactical surprise, and can take days to prepare.

The committee notes, however, that the required inventory for a leading-edge strike would certainly be far larger than for a very limited strike mission. It follows that the committee sees a potential role for CPGS for the former class of operations but is much less convinced that CTM would go far in addressing it.

Further, there is some question as to the basic requirement for a CPGS capability for a leading-edge mission. Almost by definition, many operations requiring this type of capability would have been anticipated with strategic warning and a buildup of regional forces. Moreover, the tense political environment associated with the advent of MCO would increase the risk that a strike might be construed as a nuclear attack, or as a precursor to nuclear attack. This might be especially true for a CPGS weapon delivered by a ballistic missile.

CTM in Relation to a CPGS Capability

The committee has concluded that the CPGS option represented by the CTM is best assessed in terms of a very limited strike rather than as the leading edge of something larger.[9] Many of the scenarios involving CPGS as the leading edge of MCO are likely to require the use of many prompt conventional weapons, not just the one or two missiles' worth of kinetic-energy weapons to be implemented in CTM. Although it is not at all clear that continuous global coverage would be required, a moderately larger strike might not even be possible with only two missiles per submarine launched from normal strategic patrol areas. Even if it were, each submarine that fired a CTM could temporarily expose its location. This is a more significant factor in the context of the outbreak of a major war than in the context of a very limited strike. In the long run, a CPGS capability for a leading-edge attack might be quite important, but the CTM is best suited for the very limited strike.

[9]This observation applies even more strongly to a third class of scenarios sometimes seen as supporting a CPGS capability—use of conventional weapons to cover some, but far from all, of the targets now allocated to nuclear weapons in nuclear war plans.

Finding 2: Other than forward-based systems, CTM appears to offer the only near-term option for a CPGS capability.

At least for the purpose of a very limited strike, the need for a CPGS capability could well arise before any new system can be made operational. For the next 3 years at least, the United States has no choice but to rely on existing forward-deployed forces for whatever they can provide toward a capability for CPGS. That could be a substantial contribution, because in many cases such forces can provide a quite-rapid response capability. To a considerable degree, it will be possible to identify the regions in which a need for launch is most likely, and the potentially targeted terrorist organizations would pose no air defense threat. A comprehensive policy for a prompt, if not initially literally global, strike capability should ensure that the United States has done all that is reasonable to assure that forces using conventional weapons—notably cruise missiles on aircraft, submarines, or surface ships, and attack aircraft with precision munitions (and, for some scenarios, stealth characteristics)—are appropriately deployed, trained, and supported for urgent, very limited strikes. Moreover, Congress should require that the DOD study carefully where very limited strikes are most likely over that time and how forces could be redeployed to conduct them. Furthermore, analysis and decisionmaking processes should be examined and exercised to ensure that the processes are fully understood and responsive, and that they routinely provide sound decision support reflecting likely outcomes, upside possibilities, and downside risks. If missions requiring a CPGS capability are as important as DOD has argued—and the committee agrees that they are—it is worthwhile to do whatever is necessary to fill the near-term gap as much as is reasonably possible.

In 3 years or so, however, the United States could have a new, and truly global, CPGS capability based on conversion of the Trident II (D5) missile carried aboard nuclear ballistic missile submarines (SSBNs). Some D5 re-entry vehicles (RVs) with nuclear warheads would be replaced with RVs armed with kinetic-energy weapons. These missiles would replace nuclear-armed D5s in two tubes on each SSBN. No alternative option for a CPGS capability can be made operational with comparable capabilities in so short a time. In addition to being the earliest CPGS option that could be deployed, SLBMs offer some unique advantages, chief among them the ability to remain covertly within range of potential targets for extended periods without dependence on foreign bases. The patrol locations can be chosen to allow striking most targets of interest without flying over other countries of concern.

Moreover, the committee was impressed by the deep thought and technical analysis that have gone into the Navy's proposed command and control system for conventional strike with the CTM: it would have the same extreme security and reliability as that of nuclear missiles and would be distinct in a way that would preclude the possibility of an accidental launch of a nuclear-armed Trident in a mixed load. The committee explored the matter of accidental launch and concluded that this problem has been dealt with adequately. The Navy and USSTRATCOM described safeguards that have been designed and, in most cases, integrated into the weapon system and the supporting command and control that would prevent accidental launch of the "wrong" type of missile. The committee has confidence in these measures because they are extensions of measures already used to positively prevent such an event during current operational tests.[10] The committee recognizes that other countries may not share the same high confidence with regard to proposed safeguards for preventing an accidental launch, but it is presumed that CTM would be subject to the same rigorous testing and validation procedures associated with current Trident missiles and that the proposed command and control would be sufficiently demonstrated prior to CTM becoming operational.

On the scale of such matters, the committee regards CTM development risks as low—in large part because the development would be incremental, building on the long and successful systems development history of the Navy's Strategic Systems Programs (SSP). The development and testing would resolve residual issues about terminal navigation in a complex environment; these are nontrivial, but the committee anticipates success based on the agreement to date between engineering analysis and experimentation. The challenges are much smaller than those for the longer-term systems described below.

Longer-term Options

Finding 3: Longer-term CPGS options offer potentially attractive capabilities but in some cases appear to involve high technical risk.

The committee has not yet had adequate opportunity to compare longer-term options for a CPGS capability, but its initial impression is that the Sea-Launched Global Strike Missile (SLGSM) approach presents less technical risk than the others being proposed. In perhaps 6 years or more, the SLGSM could be developed, as proposed by the Navy SSP. SLGSM is projected to carry, as one option, a large version of the bent nose lifting body already flight-tested once under conditions equivalent to those for attacking targets at ranges in excess of 2,000 nautical miles. Based on information presented

[10] Commander's evaluation tests have been conducted periodically since Polaris was initially deployed. In these tests, launches have routinely been conducted under operational conditions from submarines with mixed loads.

to the committee, that payload is designed to deliver larger munitions such as an earth penetrator for attacking deeply buried facilities. Development of the RV for the SLGSM would benefit significantly from development of the CTM RV. The committee's own analysis indicates that an SLGSM within the same volume and with a third stage added could deliver to the same range as CTM one of the smaller kinetic-energy warheads initially planned for CTM.

Less matured, longer-term CPGS delivery concepts proposed by the U.S. Army and U.S. Air Force include hypersonic boost-glide vehicles launched from the continental United States (CONUS) or forward-deployed assets and higher-speed cruise missiles launched from bombers. Although these concepts appear to have high technical risk, it has been argued that they have the potential to provide favorable system characteristics such as being less likely to be mistaken for a nuclear-armed ballistic missile, being capable of trajectory flexibility for avoiding sensitive overflights, having significant cross-range divert capability for inflight retargeting, and tailoring the end-game approach angle to a target for improved weapon system effectiveness.

The boost glide and high-speed cruise missile concepts as CPGS options require advanced technologies, especially in the areas of thermal protection and management; guidance, navigation, and control (GN&C); and submunition dispensing mechanisms. For vehicles operating in the oxidative hypersonic environment for flight times approaching an hour, advanced thermal protection and management systems will be required to insulate all internal components to below their maximum allowable temperature constraints. Advanced GN&C systems will be required for vehicles operating on global scales with portions of the trajectory potentially in GPS-denied areas. And an improved understanding of submunition dispensing mechanisms will be needed to develop high confidence in the highly dynamic processes that occur as multiple vehicles interact with each other during a high-speed deployment sequence.

The committee believes it is preferable to consider all proposed CPGS weapons as elements of a portfolio, one that needs balancing in terms of technical risk and time to deployment. The search for a single optimal system is not the best way to proceed given all of the uncertainties in the strategic environment. During the remainder of this study, the committee will complete a more detailed investigation of CPGS options together with a more complete assessment of their technology needs.

KEY RECOMMENDATIONS

At the present time, the committee provides the following key recommendation for the near-term regarding CTM, as well as one aimed at enabling attractive CPGS options for the longer term.

Recommendation 1: In FY 08, fund research, development, testing, and evaluation (RDT&E) efforts associated with CTM at a level sufficient to determine its effectiveness, but in FY 08 withhold funding for full-scale production and deployment (except any that is necessary for testing).

Although the committee has not examined the CTM program budget in detail, it is the committee's understanding that approximately $120 million in FY 08 RDT&E funds[11] would be required to maintain CTM on a near-term schedule. It believes that it is prudent to make the investment in FY 08 RDT&E needed to mature and validate the CTM capability, as well as to protect an initial operating capability (IOC) of 3 years and an option to deploy an effective CPGS capability in 3 to 5 years.

While it is not the optimal solution for the longer term, CTM offers the only viable truly global CPGS capability within the next 6 years, and it can be achieved, with military mission capabilities still to be quantified, at a relatively modest initial and life-cycle cost because of the minimal changes required in most components of the delivery system and its infrastructure. The ability of the Navy's SSP to respond to USSTRATCOM (and DOD more broadly) with this approach is possible in large part because SSP has a long history of evolutionary development of highly successful systems, which has included contractor discretionary funded work in exploring the technologies for such a capability on future SLBMs. CTM RDT&E is also a sound interim course because it provides the opportunity to address issues of military effectiveness and is a key growth path to the SLGSM, discussed above.

If a CPGS capability is desired without forward deployment in the longer term, options (other than the more mature SLGSM) presented to the committee depend on technology advances that in its judgment are more challenging and will take at least 8 years to achieve, assuming that work on those technologies is funded beginning now. Technologies developed in the CTM program should also be applicable to some CONUS-based intercontinental ballistic missile (ICBM) delivery concepts if overflight avoidance maneuvers are not required.

---

[11]CAPT Terry Benedict, USN, Technical Director, Navy Strategic Systems Programs, presentation to the committee, March 22, 2007.

Funding CTM development and end-to-end testing provides the earliest and most viable opportunity to meet the initial CPGS capability. Although there are issues about how—and indeed whether—CTM should be deployed and used that have not yet been adequately addressed, the technical feasibility of CTM has been demonstrated and the design is sound and well thought out. Accordingly, a funding path that keeps the program essentially on schedule for an IOC in 3 years and also supports the SLGSM alternative is a prudent interim step.

The committee does not, however, endorse funding for full-scale CTM production and deployment. There remain policy issues—including dealing with the ambiguity issue and consideration of alternative (albeit less-developed) systems that should be fully addressed before committing to CTM deployment. Moreover, the CTM program itself is not without technical issues that merit careful study. For example, the committee has concerns about the proposed mixed-load deployment configuration and the payload options relative to their ability to address the military needs for the target types of interest. The committee believes that alternative concepts of operation may be needed to more effectively use the capability of the system (e.g., providing larger numbers of deliverable weapons on station) while also minimizing ambiguity concerns. Given that the Trident's primary mission is nuclear deterrence, the committee also has concerns about how the CTM can be deployed most effectively for CPGS missions while avoiding crisis ambiguities.

Another example of a matter that merits continuing study relates to the bent nose payload designed for the SLGSM that has been flight-tested once under intermediate-range conditions. This payload could offer important and necessary capabilities for attacking certain existing and anticipated targets in steep terrain. The committee suggests that this configuration, beefed up for use at longer ranges, should be evaluated as part of any near-term deployed CTM capability as well. Since the tested configuration requires a thicker heat shield to reach the longer range, additional development and testing will be required to validate its performance and munitions compatibility. The committee estimates that such development and testing will require an additional year to accomplish, but emphasizes that it should be part of the ongoing CTM development effort. If successfully developed and tested, this technology would also be applicable to a three-stage SLGSM, and possibly some ICBM conventional delivery options.

In the course of CTM RDT&E (but not as a prerequisite), the Navy should investigate further some of the concerns about ambiguity, operational effectiveness, and unintended side effects on current SSBN operations. Until such issues are fully investigated, it is premature to commit to deployment or production of CTM beyond that required to build an appropriate number of full systems and conduct the tests needed to validate the design, develop different warhead capabilities, and preserve a near-term deployment option.

Recommendation 2: In FY 08, fund technology development at a level to fully support the longer-term CPGS options described in this letter report.

The committee also recommends providing a modest amount of applied research (6.2) funding toward maturing the more challenging hypersonic flight technologies needed for other longer-term CPGS options envisioned by the Air Force and the Army. This investment will permit a far sounder foundation for consideration of longer-term alternatives because it will help identify the scale of the technical challenges involved and the timelines associated with these and other alternatives.

ACKNOWLEDGMENTS

The committee thanks the following DOD officials for their help in providing and coordinating much of the committee's data-gathering during the course of the past few months: CAPT Terry Benedict, USN, Technical Director, Navy Strategic Systems Programs; Mr. James Colasacco, Division Chief, Global Strike Capability, U.S. Strategic Command; Mr. Brian Green, Deputy Assistant Secretary of Defense for Strategic Capabilities, Office of the Deputy Under Secretary of Defense (Policy); and Mr. Greg Hulcher, Director for Special Projects, Office of the Deputy Under Secretary of Defense (Acquisition, Technology, & Logistics). In addition, the committee thanks members of your staff for meeting with the committee in order to highlight some of the congressional motivations behind this study request and issues of particular concern to Congress.

We would be pleased to brief you and your staff regarding the views expressed in this letter. We remain committed to completing our final report in an expedited fashion.

Sincerely,

Albert Carnesale, *Chair*
Committee on Conventional Prompt Global Strike Capability

Enclosures:
A Statement of Task
B Committee on Conventional Prompt Global Strike Capability (as of May 2007)
C Acknowledgment of Reviewers
D Summary of Data-Gathering Sessions

cc:
Sidney Ashworth, Clerk, Subcommittee on Defense, Senate Committee on Appropriations
Charlie Huoy, Clerk, Subcommittee on Defense, Senate Committee on Appropriations

Enclosure A
Statement of Task

The conference report accompanying the FY 2007 Department of Defense Appropriations Act (H.R. 5631/Public Law 109-289) directed the National Academy of Sciences to conduct a study to analyze the mission requirement for using existing Trident II (D5) missiles with conventional payloads to provide a prompt global strike capability and, where appropriate, consider and recommend alternatives that meet the prompt global strike mission in the near term (1-2 years), mid-term (3-5 years), and the long term. The study should include analyses of the military, political, and international issues associated with each alternative. The study should consider technology options for achieving desired objectives as well as mitigating policy concerns.

This 15-month study will produce two reports: (1) a letter report following the second full committee meeting that summarizes the requirements and supporting enablers for a conventional prompt global strike capability and recommends a near-term option or options to provide this capability; and (2) a comprehensive report that addresses the full terms of reference as outlined above.

Enclosure B
Committee on Conventional Prompt Global Strike Capability (as of May 2007)

Members
Albert Carnesale, University of California, Los Angeles, *Chair*
Paul Bracken, Yale University
Paul K. Davis, The RAND Corporation
Steve Fetter, University of Maryland, College Park
John S. Foster, Jr., Rancho Palos Verdes, California
Eugene Fox, USA (Ret.), McLean, Virginia
Alec D. Gallimore, University of Michigan
Richard L. Garwin, IBM Thomas J. Watson Research Center
Eugene Habiger, USAF (Ret.), University of Georgia
David V. Kalbaugh, Centreville, Maryland
L. David Montague, Menlo Park, California
Robert B. Oakley, National Defense University
Walter B. Slocombe, Caplin & Drysdale
William D. Smith, USN (Ret.), Independent Consultant
John P. Stenbit, Oakton, Virginia
David M. Van Wie, Johns Hopkins University, Applied Physics Laboratory
Robert H. Wertheim, USN (Ret.), San Diego, California
Ellen D. Williams, University of Maryland, College Park
R. James Woolsey, Jr., Booz Allen Hamilton

Staff
Charles Draper, Director, Naval Studies Board, *Study Director*
Ian Cameron, Associate Program Officer, Naval Studies Board

Enclosure C
Acknowledgment of Reviewers

This report has been reviewed in draft form by individuals chosen for their diverse perspectives and technical expertise, in accordance with procedures approved by the National Research Council's Report Review Committee. The purpose of this independent review is to provide candid and critical comments that will assist the institution in making its published report as sound as possible and to ensure that the report meets institutional standards for objectivity, evidence, and responsiveness to the study charge. The review comments and draft manuscript remain confidential to protect the integrity of the deliberative process. We wish to thank the following individuals for their review of this report:

John F. Ahearne, Sigma Xi,
Edward G. Anderson III, USA (Ret.), Booz Allen Hamilton,
Owen R. Cote, Jr., Massachusetts Institute of Technology,
Lawrence J. Delaney, The Titan Corporation,
Sidney D. Drell, Stanford University,
Richard W. Mies, USN (Ret.), Hicks & Associates, Inc., and
Larry D. Welch, USAF (Ret.), Institute for Defense Analyses.

Although the reviewers listed above have provided many constructive comments and suggestions, they were not asked to endorse the conclusions or recommendations nor did they see the final draft of the report before its release. The review of this report was overseen by Stephen Berry of the University of Chicago. Appointed by the National Research Council, he was responsible for making certain that an independent examination of this report was carried out in accordance with institutional procedures and that all review comments were carefully considered. Responsibility for the final content of this report rests entirely with the authoring committee and the institution.

Enclosure D
Summary of Data-Gathering Sessions

The Committee on Conventional Prompt Global Strike Capability first convened in February 2007 and held two full committee meetings and two subcommittee meetings prior to issuing this letter report. In addition to deliberating on and preparing its letter report, the committee also participated in the data-gathering sessions at these meetings, which are summarized below.

- *February 22-23, 2007, in Washington, D.C.* First full committee meeting: Briefings on policy, requirements, supporting enablers, and technology plans for conventional prompt global strike from the Office of the Under Secretary of Defense for Policy; U.S. Strategic Command; Office of the Under Secretary of Defense for Acquisition, Technology, and Logistics; U.S. Navy Strategic Systems Programs; U.S. Air Force Space Command; and Defense Intelligence Agency.
- *March 15, 2007, in Washington, D.C.* First subcommittee meeting (a makeup of the first full committee meeting for members not in attendance): Briefings on policy, requirements, supporting enablers, and technology plans for conventional prompt global strike from the Office of the Under Secretary of Defense for Policy; U.S. Strategic Command; Office of the Under Secretary of Defense for Acquisition, Technology, and Logistics; U.S. Navy Strategic Systems Programs; and U.S. Air Force Space Command.
- *March 16, 2007, in Washington, D.C.* Second subcommittee meeting. Briefings on intelligence capabilities for conventional prompt global strike from the Office of the Under Secretary of Defense for Policy and U.S. Strategic Command.
- *March 22-23, 2007, in Washington, D.C.* Second full committee meeting: Briefings on short-, mid-, and long-term options for conventional prompt global strike, as well as policy and technical concerns associated with each, from congressional staff, U.S. Senate Defense Appropriations Subcommittee; Congressional Research Service; U.S. Strategic Command; U.S. Air Force Space Command; U.S. Navy Strategic Systems Programs; Office of the Under Secretary of Defense for Policy; and U.S. Army Space and Missile Defense Technical Center. In addition, Dr. Pavel Podvig, Center for International Security and Cooperation, Stanford University; Dr. Theodore Postol, Security Studies Program, Massachusetts Institute of Technology; and Dr. Jeffrey Lewis, New America Foundation, provided in a data-gathering session open to the public their views on international security, arms control, and technical issues related to conventional prompt global strike.

F

Summary of Committee Meeting Agendas

The Committee on Conventional Prompt Global Strike Capability first convened in February 2007 and held additional meetings and site visits over a period of 8 months, as summarized below:

- *February 22-23, 2007, in Washington, D.C.* First full committee meeting: Briefings on policy, requirements, supporting enablers, and technology plans for conventional prompt global strike from the Office of the Under Secretary of Defense for Policy; U.S. Strategic Command; Office of the Under Secretary of Defense for Acquisition, Technology and Logistics; U.S. Navy Strategic Systems Programs; U.S. Air Force Space Command; and Defense Intelligence Agency.
- *March 15, 2007, in Washington, D.C.* First subcommittee meeting (a makeup of the first full committee meeting for members not in attendance): Briefings on policy, requirements, supporting enablers, and technology plans for conventional prompt global strike from the Office of the Under Secretary of Defense for Policy; U.S. Strategic Command; Office of the Under Secretary of Defense for Acquisition, Technology and Logistics; U.S. Navy Strategic Systems Programs; and U.S. Air Force Space Command.
- *March 16, 2007, in Washington, D.C.* Second subcommittee meeting: Briefings on intelligence capabilities for conventional prompt global strike from the Office of the Under Secretary of Defense for Policy and U.S. Strategic Command.
- *March 22-23, 2007, in Washington, D.C.* Second full committee meeting: Briefings on short-, mid-, and long-term options for conventional prompt global strike, as well as policy and technical concerns associated with each, from congressional staff, U.S. Senate Defense Appropriations Subcommittee; Congres-

sional Research Service; U.S. Strategic Command; U.S. Air Force Space Command; U.S. Navy Strategic Systems Programs; Office of the Under Secretary of Defense for Policy; and U.S. Army Space and Missile Defense Technical Center. In addition, Dr. Pavel Podvig, Center for International Security and Cooperation, Stanford University; Dr. Theodore Postol, Security Studies Program, Massachusetts Institute of Technology; and Dr. Jeffrey Lewis, New America Foundation, provided in a data-gathering session open to the public their views on international security, arms control, and technical issues related to conventional prompt global strike.

- *May 10-11, 2007, in Washington, D.C.* Third full committee meeting: Briefings on deterrence aspects, treaty implications, policy and operational perspectives, and geospatial intelligence from the Department of Defense, Department of State, and National Geospatial-Intelligence Agency.
- *June 12-13, 2007, in San Diego, California.* Fourth full committee meeting: Briefings on requirements and doctrine, military unity, and submarine operations from the Defense Threat Reduction Agency, Department of Defense, U.S. Strategic Command, and U.S. Pacific Fleet.
- *July 14-15, 2007, in San Diego, California.* Fifth full committee meeting: Briefings on the Conventional Trident Modification and weapon-on-target effectiveness from the U.S. Strategic Command, U.S. Air Force, U.S. Navy, Defense Threat Reduction Agency, Sandia National Laboratories, and Office of the Secretary of Defense.
- *August 9-10, 2007, in Washington, D.C.* Sixth full committee meeting: Briefings from the Department of Defense on scenarios for conventional prompt global strike and briefings from industry (Alliant Techsystems, Boeing Company, Lockheed Martin Corporation, Northrop Grumman Mission Systems, and Raytheon Corporation) on potential conventional prompt global strike alternatives.
- *September 17-21, 2007, in Irvine, California.* Seventh full committee meeting: Committee deliberations and report drafting.

G

The Why and How of Boost-Glide Systems

Given the prominence of the boost-glide technology in some of the options under consideration in this report, it is useful to include an appendix explaining semiquantitatively what the technology can and cannot accomplish, its relation to the fractional orbit bombardment systems (FOBSs) technology discussed during the 1960s and 1970s, and some of the technical challenges involved. Another issue is the extent to which such vehicles can be expected to defeat "garden-variety" and advanced air defenses.

A boost-glide vehicle (BGV), or "lifting body" without propulsion, can be used to extend the range of a ballistic-missile payload beyond the purely ballistic range. It can also be used for out-of-plane or "dogleg" maneuvers to avoid overflight of certain areas or to allow the dropping of initial rocket stages into the sea or into another body of water not under the ballistic path. The space shuttle on reentry is an example of a hypersonic lifting body.

First consider the BGV for relatively short-range systems—up to a few thousand kilometers in range—in the approximation of a flat Earth. An important simplification arises from the fact that the atmosphere is shallow; the air density falls by a factor $e = 2.72$ for each 8 km increase in altitude. As is the case with a normal glider, the aerodynamic support of the vehicle against gravity (the "lift" L) is accompanied by "drag due to lift," as characterized by the lift-to-drag ratio (L/D); for a clean subsonic glider aircraft this may be as much as 40, but for a hypersonic lifting body an $L/D = 2.2$ is an achievement. In the numerical examples, it is assumed that $L/D = 2.2$.

With the glider aircraft or for a powered vehicle that has run out of fuel and that is gliding for as long a distance as possible, the drag, D, extracts energy from the vehicle—from its store of kinetic energy $MV^2/2$ and potential energy MgH;

here *M, V, g,* and *H* are the vehicle mass in kilograms, the velocity in meters per second, the gravitational acceleration 0.0098 kilometers per second, and the flight altitude in kilometers. Every pilot has first in mind the glide range from altitude, which by the same token (equating the loss of potential and kinetic energy to the drag times the distance) is exactly *L/D* multiplied by $(H + (MV^2/2g))$. An airliner with *L/D* about 20 can glide 20 times the initial altitude, which is quite significant—about 200 km from an initial altitude of 10 km. Added to this is the contribution of initial kinetic energy, corresponding to an additional altitude of 4.5 km at Mach 1, about 300 m/s. For the private pilot, the kinetic energy term is not large, since small aircraft may travel at 0.3 Mach, so that the equivalent height is only about 0.5 km, and glide range comes mostly from altitude. The relative importance of speed and altitude is very different for hypersonic speed, since Mach 20 would equate to $(20)^2 \times 4.5$ km, or 1,800 km altitude.

Consider the use of glide for range extension of the maximum-range ("minimum energy") ballistic trajectory. A pure ballistic trajectory to intercontinental range has the reentry vehicle (RV) reentering the atmosphere near the target at a grazing angle typically on the order of –22° and slowing abruptly in the atmosphere according to the M/C_dA of the RV, with *M* the RV mass, C_d the drag coefficient, and *A* the base area of the RV. This ballistic reentry wastes the kinetic energy of the RV at the time of reentry, whose velocity is the minimum required to achieve the desired range in the first place. This is the best that could be done if Earth had no atmosphere. But it does, and in principle the RV could be designed as a lifting body for the hypersonic regime, and if the thermal insult could be managed it could transition in the upper regions of the atmosphere to near-horizontal flight, and then use lift and change of altitude, air density, and change of angle of attack to support the RV weight for a substantial range extension beyond the purely ballistic trajectory. This approach was validated decades ago by flights of the Mk-500 "Evader" RV.

Successful implementation of boost-glide technology could yield additional benefits for the prompt global strike mission by means of the ability to maneuver and thus to aid in avoiding undesired overflight of various countries. The launch would be similar to that for a minimum-energy trajectory—that is, maximum range for a given missile—typically with a high apogee and the transition on ballistic reentry to either level or phugoid (porpoise-like) flight—in which the RV bounces in and out of the atmosphere several times and supports its weight by aerodynamic lift only a relatively small fraction of the time, say 10 percent. Supporters of the BGV often argue that this phugoid flight provides range extension at little cost, because for much of this flight—between bounces—the drag is almost zero.

It is important to recognize that there is "no free lunch" in phugoid flight, because the lift averaged over this portion of the flight is precisely the weight of the vehicle, and so the time-average lift (and drag) are the same as if the RV were flying at steady altitude and speed in order to maintain the same average aerodynamic lift. The average lift must be equal to the weight of the vehicle: $W = gM$; the average drag is thus the weight divided by (L/D).

On the assumption of constant L/D, it turns out that there are simple closed-form formulas not only for the glide portion of flight but also for the velocity and kinetic-energy loss in the transition from ballistic flight to glide.

FLAT EARTH

1. On a flat Earth, the maximum ballistic range is achieved with a constant launch angle of 45° to the horizontal; the purely ballistic range is $R_b = V^2/g$, as is readily derived. Here V is the initial speed of the projectile, T the time of ballistic flight, g the acceleration of gravity (9.8 m/s^2), and R_b the resulting ballistic range:

$$V_h = V_v = V/\sqrt{2};\ \mathrm{T} = 2V_v/g;\ R_b = V_h \times T = V^2/g.$$

V_h and V_v are the horizontal and vertical components of V.

2. Let s be the horizontal path length at a given time. It is shown that the glide range R_g with constant L/D is half the ballistic range multiplied by L/D, and so the glide range equals the ballistic range for $L/D = 2$:

$$dV/ds = dV/dt \times dt/ds = (g/(L/D)) \times (1/V);\ VdV = (g/(L/D))ds;$$

Integrating, $V^2/2 = gR_g/(L/D)$, so $R_g = (L/D) \times V^2/2g =$
$(L/D) \times R_b/2$ (Eq. G-1)

The glide range can thus significantly extend the ballistic range, but not to the full extent implied here. The launch velocity (and reentry velocity) cannot be taken as the initial glide speed because of the loss of speed due to lift-induced drag in the maneuver to glide flight.

3. Although it is probably not optimum, the transition between ballistic reentry and horizontal glide can be assumed to be made with a high-g ($g' >> g$) trajectory of constant radius, r, and with the assumption of constant L/D the logarithmic fraction of velocity decrement is just the angle of the arc, θ (in this case 45°) divided by (L/D):

$dV/d\theta = dt/d\theta \times dV/dt = (r/V) \times dV/dt = (r/V)(D/M) = (r/V)\ g'/(L/D)$, with $g' = V^2/r$ the centripetal acceleration.

So $dV/d\theta = V/(L/D)$, from which $dV/V = d\theta/(L/D)$.

Integrating, $\Delta \ln V = (D/L)\Delta\theta$, so that

$$(V_f/V_i) = exp((D/L) \times \Delta\theta) = exp -((1/2.2) \times (\pi/4)), \quad \text{(Eq. G-2)}$$

where V_i is the vehicle initial speed at the beginning of the transition, and V_f is the speed at the end of the transition and the beginning of the glide portion of flight.

For $\Delta\theta = 45°$ ($= \pi/4$) and $(D/L) = 1/2.2$, the velocity falls to 0.6998 of its initial value, and the square of the velocity to 0.4897. The glide portion of the trajectory, R_b, is thus reduced to 49 percent of what it would have been had the projectile been fired horizontally in the first place into glide mode at $V = V_i$ and at constant L/D = 2.2. The ballistic-trajectory range extension is thus 53.9 percent of the ballistic range, rather than the 110 percent if the vehicle could have negotiated the –45° pull-up without velocity loss on a flat Earth or at short range, as would be the case in principle if a long tether could have been used to supply the centripetal force for the maneuver.

At intercontinental ballistic missile (ICBM) range on a round Earth with a reentry dip angle of –22.5°, the logarithmic loss of velocity is only half as large as at –45° (the velocity emerging from the maneuver is 0.8365 of the reentry speed, and its square 0.700).

CALCULATION OF TRAJECTORY FOR A ROUND, NONROTATING EARTH

As indicated in Figure G-1, much of the range benefit from boost-glide in general and phugoid flight in particular is only available on the round Earth and with near-orbital initial speed of the RV. For speeds in the upper atmosphere comparable to the orbital velocity in low Earth orbit (LEO), almost no aerodynamic lift is necessary, so the glide range can be astonishing—say, on the order of 13,000 nautical miles (nmi) in some cases. Since Earth's circumference is 360 degrees of arc and each arc minute is 1 nmi, the circumference of the world is 21,600 nmi (precisely 40,000 km, by the definition of the meter).

For simplicity, one can calculate numerically the time-reversed trajectory, starting at zero speed over the target with an (unphysical) angle of attack that is assumed to support the weight of the vehicle (gM) at these very low speeds. Since the weight of the vehicle, W, is constant (it does not have fuel or propulsion), the drag is also constant at $W/(L/D)$, and thus the horizontal acceleration of the vehicle is constant and known: $D/M = g/(L/D)$.

The calculation is done in an Excel spreadsheet, providing at each step the new velocity and the integral of velocity thus far, corresponding to the range from the assumed target.

A plot of velocity versus range, for example, Figure G-2, shows that centrifugal force provides 64 percent of the overall lift of the vehicle with a range of 7,168 km (3,870 nmi) to go, and a time-to-go of 1,950 s. Supplementing the curve of velocity versus range of Figure G-2 are Figure G-3, "Range versus time

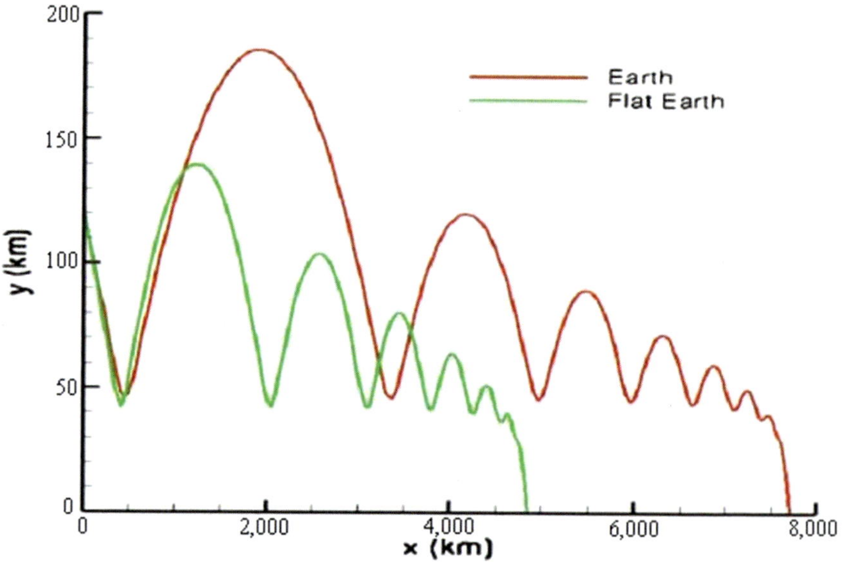

FIGURE G-1 Reentry trajectories for *L/D* = 2.2. Note that the last 1,000 km or more of the reentry trajectories are identical for Earth and flat Earth. SOURCE: Data for initial conditions (7 km/s, –10° grazing) provided by G. Candler, University of Minnesota, personal communication to the committee, September 17, 2007.

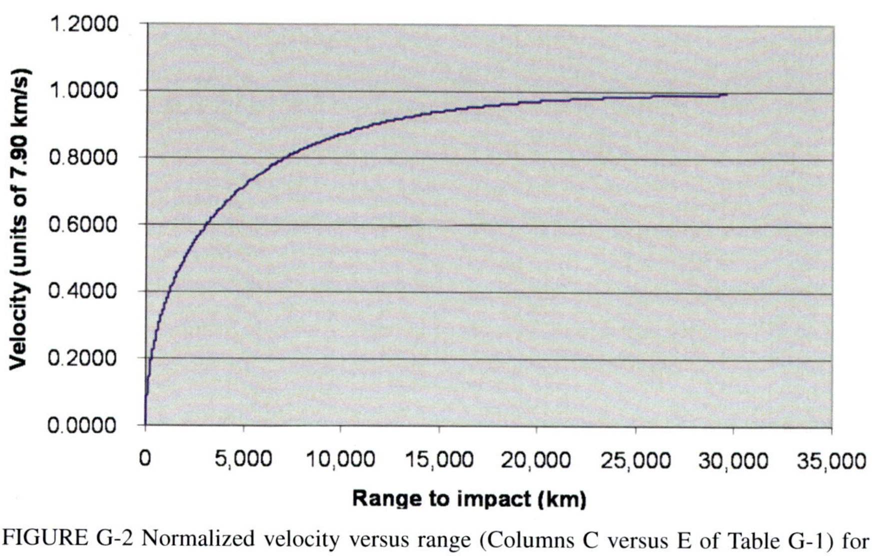

FIGURE G-2 Normalized velocity versus range (Columns C versus E of Table G-1) for pure glide.

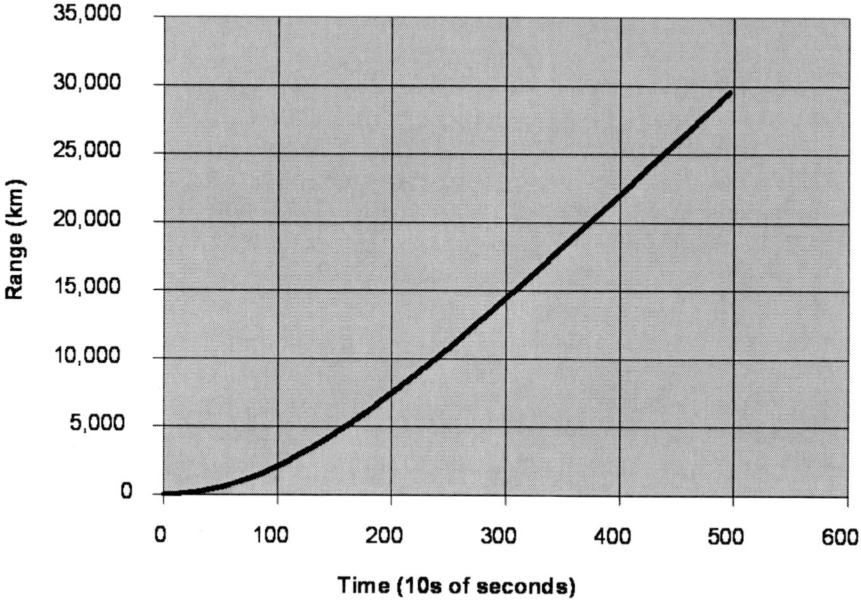

FIGURE G-3 Range versus time for pure glide (Columns E versus B of Table G-1). Note that the unit of time here is 10 s.

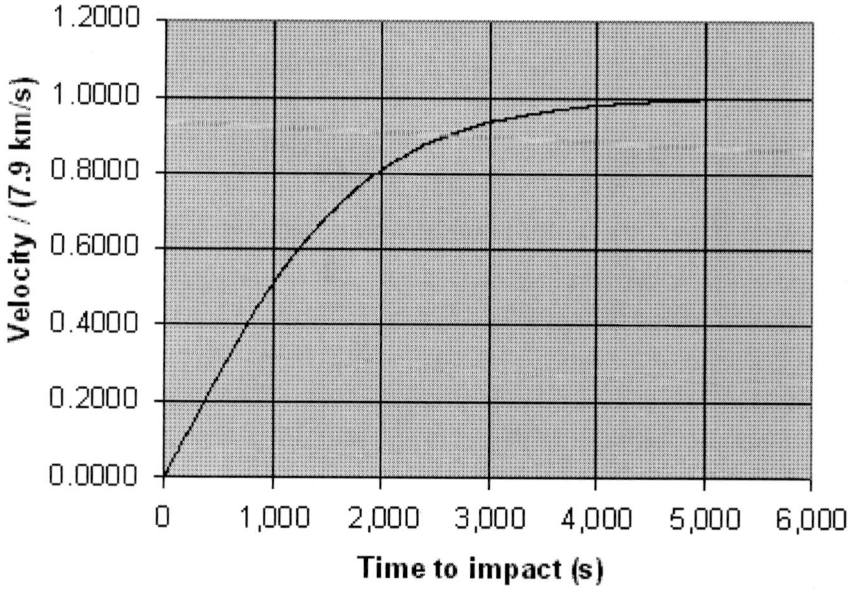

FIGURE G-4 Velocity versus time for pure glide (Columns C versus B of Table G-1).

for pure glide," and Figure G-4 (see previous page), "Velocity versus time for pure glide," for cases in which time is of interest.

The spreadsheet, of which excerpts are shown in Table G-1, embodies calculation results using the following formulas: beginning at rest, $dV/dt = g\ D/L$; at high speeds and constant altitude, aerodynamic lift does not need to compensate g—only $g \times (1 - (V/Ve)^2)$, where Ve is LEO speed of 7.90 km/s. Using the prime to indicate time derivative,

$V' = g\ (D/L)\ (1 - (V/Ve)^2)$, or setting $v = V'/Ve$, we have

$v'/(1 - v^2) = (g/Ve)\ (D/\mathrm{L}) = c$,
defining c as $(\mathrm{g}/Ve)\ (D/L) = 0.000563$ for $L/D = 2.2$. (Eq. G-3)

The scaled speed, v, is obtained by accumulating v' ("v-dot" in the spreadsheet), and the range-to-date by accumulating v.

The following simple closed-form integration has been obtained:

$v(t) = \sin(-\pi/2 + 2\ \mathrm{atan}(\exp(ct)))$ (Eq. G-4)

In Equation G-3, substitute $v = \sin(\varphi)$, so that $v' = \varphi' \times \cos(\varphi) = c$, and

$(d\varphi \times \cos(\varphi)/((\cos(\varphi))^2) = c\ dt$, or $d\varphi/(\cos(\varphi)) = c\ dt$.

This integrates to $\log \tan(\pi/4 + \varphi/2) = ct$, and $\tan(\pi/4 + \varphi/2) = \exp(ct)$, so that

$\varphi = -(\pi/2) + 2\ \mathrm{atan}(\exp(ct))$. Substituting v for $\sin(\varphi)$, $v(t) = \sin(\varphi) = \sin(-\pi/2 + 2\ \mathrm{atan}(\exp(ct)))$.

Spreadsheet column D, headed "v-analytic" (Table G-1) represents values of this last formula, for comparison with the column C, "v," resulting from numerical integration of v-dot. The agreement to about 0.1 percent is good for a simple first-order integration but, more importantly, shows the absence of blunders in either calculation.

COMPARISON WITH TRADITIONAL CALCULATION OF PHUGOID GLIDE

Figure G-1 shows glide range for a flat Earth of 4,800 km for the assumed initial conditions, and 7,700 km for the round Earth. According to Equation G-1, for a flat Earth the glide range is $s = (L/D) \times V^2/2g$, which for $V = 6.5$ km/s amounts to 4,742 km. Why "6.5 km/s"? Because 7 km/s at –10° grazing angle yields $V_o/V_i = e^{-(D/L)\ \Delta\theta}$ or 0.93, so that a $V_i = 7.0$ km/s becomes $V_o = 7.0 \times 0.93 = 6.5$ km/s

TABLE G-1 Excerpts of Calculation Results from Spreadsheet

	A	B	C	D	E	F	G	H	I	J
1	Back-integration of glide vehicle. 08/09/2007					"08-09-2007 Boost-Glide (FOBS)1.xls"				
2				L/D = 2.2	Ve=7.90 km/s g=0.0098 km/s^2			c=(g/Ve) (D/L)=		0.000563
3	v-prime	t	v	v-analytic	R		V-prime	t	V	
4	5.63E-04	0	0.0000	0.0000	0	A4=J2*(1-C4*C4)	0.0044	0	0.004	
5	5.63E-04	10	0.0056	0.0056	0	B4=0; C4=0; E4=0;	0.0044	10	0.044	
6	5.63E-04	20	0.0113	0.0113	1	D4=SIN((-PI()/2)+2* ATAN(EXP(J2*$B4)))	0.0044	20	0.089	
7	5.63E-04	30	0.0169	0.0169	2		0.0044	30	0.133	
8	5.63E-04	40	0.0225	0.0225	4	H4=$B4	0.0044	40	0.178	
9	5.63E-04	50	0.0281	0.0281	6	I4=$C4*7.9	0.0044	50	0.222	
10	5.62E-04	60	0.0338	0.0338	8	B5=B4+10	0.0044	60	0.267	
11	5.62E-04	70	0.0394	0.0394	11	C5=C4+10*A4	0.0044	70	0.311	
12	5.62E-04	80	0.0450	0.0450	14		0.0044	80	0.356	
13	5.62E-04	90	0.0506	0.0506	18		0.0044	90	0.400	
14	5.61E-04	100	0.0562	0.0562	22		0.0044	100	0.444	
15	5.61E-04	110	0.0619	0.0619	27		0.0044	110	0.489	
141	3.26E-04	1370	0.6486	0.6477	3820		2.58E-03	1370	5.124	
142	3.24E-04	1380	0.6518	0.6509	3871		2.56E-03	1380	5.150	
143	3.21E-04	1390	0.6551	0.6542	3923		2.54E-03	1390	5.175	
144	3.19E-04	1400	0.6583	0.6574	3975		2.52E-03	1400	5.201	
145	3.17E-04	1410	0.6615	0.6606	4027		2.50E-03	1410	5.226	
804	2.64E-07	8000	0.9998	0.9998	53500		2.09E-06	8000	7.898	
805	2.61E-07	8010	0.9998	0.9998	53579		2.06E-06	8010	7.898	
806	2.58E-07	8020	0.9998	0.9998	53658		2.04E-06	8020	7.898	
807	2.55E-07	8030	0.9998	0.9998	53737		2.02E-06	8030	7.898	

NOTE: Columns A, C, and D are normalized to 7.90 km/s. Columns G and I are km/s^2 and km/s.

after the pull-up of 10° to the horizontal. The comparison between the 4,800 km estimated from the curve in Figure G-1 and the average-lift approximation result of 4,742 km is good.

From Figure G-1, the round-Earth glide range can be estimated as 7,700 km. One can interpolate for an initial glide speed of 6.5 km/s to find a glide range of about 7,900 km—again in reasonable agreement with the result of the detailed calculation as displayed in Figure G-1.

Note that entry into glide flight horizontally at 7 km/s, rather than at –10°, would provide a glide range of 10,740 km; most of the increase arises from the closer approach to orbital speed of 7.90 km/s, with the resultant reduction in needed aerodynamic lift and hence drag.[1]

Some of the proposals for long-range boost-glide vehicles enter the glide phase at angles from the horizontal two to four times smaller than the 10° example used here, and at speeds considerably closer to orbital speed of 7.90 km/s (25,920 ft/s) than the example of 7 km/s used here. They are essentially "fractional orbital bombardment vehicles" with essentially infinite "range extension" and substantial cross-range maneuver capability.

INTERPRETATION

It is beyond the scope of the appendix to analyze quantitatively the major challenge to the BGV—the long duration of the heat influx into the thermal protection system that shields the structure and internals of the vehicle from the fiery heat of skin friction with Earth's atmosphere. BGVs of longest range start at essentially orbital speed of 7.9 km/s and thus have more kinetic energy to dissipate than do ICBM RVs. The RV, however, traverses the 8 km "scale height" of the atmosphere at an angle to the horizontal of 22°, in a few seconds, while the BGV supports itself aerodynamically for 10,000 km at near-orbital speed for 1,200 s. The heating due to lift is concentrated on the lower surface of the BGV rather than uniformly around the axis of the RV, usually resulting in a very thick layer of ablative material on the lower surface of the BGV. The function of this inner layer is simple insulation rather than ablation, and so the thermal protection

[1]At steeper reentry angles than the 10° example of Figure G-1, the agreement of detailed calculation with the time-averaged lift and drag calculated here is much worse, because the approximation of horizontal flight is increasingly violated with the steeper angles of the phugoid flight pictured in Figure G-1. For instance, flat-Earth glide with 45° pull-up would give 2,693 km range extension, while detailed calculation (G. Candler, University of Minnesota, personal communication to the committee, September 19, 2007) of conventional phugoid flight provides only 1,790 km. Taking into account that the detailed calculation typified in Figure G-1 does not change the horizontal component of ballistic reentry velocity, this fraction 0.707 would reduce the time-averaged drag range to 1,904 km, for comparison with the detailed calculation of 1,790 km. This shows the merit of nearly horizontal injection at the altitude that will provide $L/D = 2.2$ at the injection speed, so that long-distance glide can be achieved by a gradual drop of altitude in order to provide steady 1-g lift as the vehicle speed gradually drops.

system has a different optimum design than does that of an RV. Indeed, much of the protection system could be in the form of non-ablating material such as the "tiles" on the space shuttle.

The intense heating of the BGV during the whole of the glide phase provides a strong infrared signal to defensive systems equipped to detect it or to use it for an infrared homing intercept.

A simple terminal maneuver for a ballistic missile will allow it to deny sanctuary to structures and locations shielded by a near-vertical bluff. At intermediate range this can require a 45° maneuver that with an $L/D = 2.2$ would (according the example following Equation G-2) result in a reduction of warhead speed to 0.6998 of the initial speed. If performed at 10 *g* transverse acceleration (0.098 km/s^2), the maneuver could take on the order of 30 s; an alternative would be to have a high-drag RV to greatly reduce speed to, say, Mach 3 (1 km/s), so that a 45° maneuver could be accomplished in a few seconds (slowdown to turn). The simple kinematic considerations of this appendix indicate the value of the engineering design of a variable-geometry RV, and the competition between the longer-term "better" and the earlier and perhaps "good enough."

H

Cooperative Reduction of Nuclear Ambiguity

A major concern expressed in regard to Conventional Trident Modification (CTM) and other conventional prompt global strike (CPGS) options is that a vehicle delivering a non-nuclear warhead might be mistaken for one delivering a nuclear warhead, heightening the risk of a nuclear response. As discussed elsewhere in this report, few states in the near future will be capable of detecting the launch and direction of a missile launched from a submarine, and relatively few will be capable and concerned about a U.S. silo-based missile launch.

Nevertheless, there are powerful tools available on a bilateral basis for communicating and validating an assertion that a delivery vehicle in flight is not carrying a nuclear warhead. One mature set of tools would be an authenticated freeze-frame video surveillance camera suitable for installation in a silo (or submarine) to ensure that the missile front end has not been changed after mutual inspection. Such video cameras are in wide use in verifying agreements on nuclear materials, both bilateral and under the International Atomic Energy Agency, transmitting encrypted and validated pictures over the Internet.

The Strategic Arms Reduction Treaty (START) of 1991 limits the United States and Russia each to 6,000 "accountable" strategic nuclear warheads with no more than 4,900 of them delivered by ballistic missiles. Furthermore, START created protocols for inspection to ensure that each missile had no more than its allowable number of nuclear warheads—a more difficult task than demonstrating that a missile is not carrying even one nuclear warhead. These measures have long been in place between the United States and Russia, without significant problems. The baseline inspections began March 1, 1995, and were completed in 120 days.

To complete the bilateral ambiguity-mitigation system would require the United States to inform its bilateral partners from *which* silo the conventional missile was just fired; if this silo was one of those subject to the proffered inspection regime, Russia would know that it did not contain a nuclear warhead. Russia would probably be able to verify the launch to a particular silo by its own information sources. However, the preferred approach is precisely parallel to that proposed below for the submarine missile launch, in which the video camera and its communication channel transmit information that a conventionally armed missile was launched.

Other nations might be similarly assured if such bilateral means were extended to them (even if not reciprocated) or if these means were internationalized.

MITIGATION OF NUCLEAR AMBIGUITY FROM SUBMARINE MISSILE LAUNCH

There is a certain economy of argument and even implementation if submarine-launch ambiguity is abated in the same cooperative fashion as that for silo-launch. Complications arise, however, in that the United States does not want to reveal routinely the location of its submarines, and also in that different tubes on a submarine cannot be distinguished by geographic location, as they can in the case of silos. Nevertheless, a similar approach can be used. Here, a video camera would be mounted within the missile tube, observing the missile shroud and sending its authenticated, encrypted pictures to the submarine for handling.[1] Included in the picture or signal is a clock feature to be provided by Russia, to time-stamp each picture, as well as an indication of launch, such as a pressure transducer to record the pressure rise in the tube as the gas generator expels the missile.

Pictures would be recorded every hour or so and reviewed every few days routinely in the submarine, or, more rarely, on request relayed to the bilateral partner. At the time of launch, the most recent time-stamped pictures would be transmitted immediately from the submarine, together with the final picture showing pressure-pulse or other evidence of missile launch; the transmission would be received through U.S. Navy means and relayed instantly from the United States to Russia. The launch evidence assures Russia (and other possible bilateral partners) that the particular non-nuclear missile was launched, and avoids the concern regarding a hoax of transmitting a reassurance picture accompanied by the launch of a nuclear-armed missile from a different launch tube. Because, the launch of

[1]While a submarine-launched ballistic-missile tube has considerable room at its base, it has very little spare room around the missile itself. However, a digital imager with volume of a cubic centimeter and coupled with lightweight plastic mirrors and light-emitting diodes could provide the necessary illumination and field for viewing. An alternative approach might be to mount the verification video camera inside the missile shroud, near the mounting fixture of the reentry vehicle, although this would require detailed analysis to show that the video camera would not interfere with flight operations. This alternative approach might allow for the use of a graphic integrating accelerometer, such as a spring-constrained ball in a transparent cylinder of silicone fluid that would move as the missile emerged from the tube, thus indicating a launch.

a submarine-launched ballistic missile (SLBM) produces a "pillar of smoke by day" or "a pillar of fire by night," there should be no U.S. reluctance to transmit this final verification picture at the time of launch.

There might be concern that the conventionally armed Trident might be ejected (and the ejection authenticated as described here) but without the ignition of its rocket motor; some tens of seconds later a nuclear-armed Trident might be launched from another tube of the same SLBM, and mistaken by Russian reconnaissance systems for the conventionally armed Trident. This fraud could be precluded by the authenticated transmission, 1 minute after the launch-evidence transmission, of a 2-minute record from a hull-mounted accelerometer in the SLBM that would show the characteristic signal of a single missile launch or, alternatively, the signal of two launches.

MITIGATION OF NUCLEAR AMBIGUITY FROM SILO-BASED MISSILE LAUNCH

There is vastly more room in a silo than in a submarine missile tube, so the mounting of the video camera is a much simpler task in the former, and there is no inhibition to near-continous transmission of the authenticated picture. Again, the indication of launch could be the pressure pulse as the missile was expelled from the silo. As in the submarine-launch case, the video camera could alternatively be mounted within the missile shroud, with a picture of the graphic integrating accelerometer as proof of launch.

CONCLUSION

It is clear that cooperative technical measures can be implemented at low cost and low risk, and, even more so, that these same measures would alleviate nuclear ambiguity in the event that a non-nuclear warhead was launched from a U.S. silo or submarine. Of course an engineering study needs to be done to choose the appropriate encrypted, validated camera approach for this cooperative mitigation of ambiguity.

It is not the purpose of this appendix to advocate the adoption of the cooperative measures described here. As described in the main text of the report, the committee concluded that the ambiguity problems could be managed with low risk. However, for readers who disagree and believe that ambiguity is a major concern, this appendix outlines a feasible technical solution that depends only on a low-risk extension of proven technology. For those who doubt the utility of this solution, it should be noted again—as in the main text—that a nuclear weapon could be delivered by any long-range aircraft or missile, even if that aircraft or missile had not previously been associated with nuclear weapons. That would be easier than finding a way to trick the technical safeguard described above. The fundamental ambiguity problem of CPGS would not be resolved by adopting a strategic delivery platform other than the Trident.

I

The Minuteman Option

Early in the course of this study, the Committee on Conventional Prompt Global Strike Capability was indeed puzzled that the Department of Defense's (DOD's) list of actively considered options did not include putting a conventional warhead on the Minuteman III missile, since those missiles and operational procedures exist, are well understood, and have significant payload. Moreover, various possible applications for the missiles have been suggested over the years by scientific panels; conventional global strike seemed to be another natural application to consider. Upon further inquiry, however, the committee independently concluded that for reasons described below, the Minuteman option is not as attractive as the sea-based alternatives for the near term, and that it is not as attractive as a number of the longer-term options.

The salient points are best read sequentially:

- The Strategic Arms Reduction Treaty (START) of 1991 may well be extended, in which case using the Minuteman for conventional prompt global strike (CPGS) would probably require treaty modifications that would have to be negotiated with Russia and other countries, raising the possibility of complications, delays, and uncertainties.
- If treaty issues did not exist, the Minuteman launched from current installations in California or Florida would often have to overfly Europe, Russia, or China, creating nontrivial risks of international incidents and possible ambiguities during crisis.
- Deploying the Minuteman elsewhere, such as in Hawaii, which the committee analyzed, could largely avoid the overflight problems, but the political, economic, and procedural difficulties of introducing a new and secure operational

base for strategic missiles, even if the missiles are few in number and conventionally armed, might prove considerable and cause delays.

- Even if the basing difficulties could be resolved quickly, the Minuteman option proves not to have any particular attractions relative to the submarine-based approaches. Indeed, it would probably have to be modified to use something very similar to the Conventional Trident Modification's (CTM's) front end in order to achieve adequate accuracy and to build conservatively on past technology. Moreover, such a modification would take time and testing and would incur the difficulties that arise when one organization is asked to use technology developed by another (e.g., technology tested in a Defense Advanced Research Projects Agency [DARPA] program). The difficulties are not merely parochial, but also technical: organizations have rich infrastructures developed over decades, which include people with a great deal of tacit knowledge that must also somehow be captured and transferred. The Air Force might find it necessary or expedient to develop its own version of a CTM-like front end, in which case additional years might be required. Further, development risk would be higher than in the Navy's incremental approach within a mature organization that has a long-term record of success with almost precisely the same components as in the CTM or CTM-2.
- The committee reasoned, nonetheless, that if concerns about ambiguity problems in time of crisis were a dominant consideration, then perhaps the Minuteman option would nonetheless have some advantages. It should be possible to provide on-site inspections and persistent instrumentation that would provide nations such as Russia with a reasonably high level of assurance that any launches would contain conventional weapons ("reasonably high" because nearly any system of assurances could probably be defeated with sufficient cunning and lapses of diligence by the observer). However, the committee's analysis concluded that with only modestly more creative measures, such assurances could be provided with submarine-based options (see Appendix H).

For all of these reasons, then, as well as the fact that the DOD had earlier considered the option and rejected it, the committee did not analyze the Minuteman option to the same extent that it did the others.